soma basics

lighting

somabasics lighting

Sebastian Conran & Mark Bond

special photography by Thomas Stewart

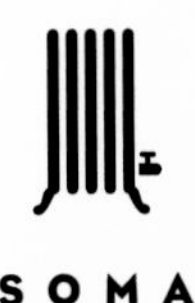

SOMA

To Sam, Max & Lola—light of our lives

contents

And God said, Let there be light: **and there was light.**

And God saw the light, that it was good: and God divided the light from the darkness.

And God called the light **Day**, and the darkness he called **Night.** And the evening and the morning were the first day.

introduction

There is more to light than meets the eye.

The Bible is crammed with positive references to light, as well as negative ones to darkness. There are also indications in many other ancient texts that our forebears had a strong attitude toward and spiritual relationship with light. These days, we know that, through photosynthesis, it is light that in fact gives life . . . well, if it does not exactly give life, it does allow life to continue. From a more secular, scientific perspective, light is earth's principal source of natural, renewable energy, from which all life stems. Without it, plant life and therefore animal life could not survive.

Light travels at 186,000 miles per second, the fastest attainable speed yet known to man. It is also the scientific absolute that relates to time. Whether you are interpreting Albert Einstein's theories of relativity or Stephen Hawking's "light cones," there does not appear to be anything beyond the speed of light. In fact, as far as I can fathom from their complex arguments, if you were to travel faster than light you would theoretically be going back in time—or whatever else hyperspace promises—light-years from now.

Light communicates our surroundings to us—this is the familiar visible light that bounces off objects into our eyes. Natural daylight from the sun consists of a cocktail of colored wavelengths that blend together in

Our design studio was originally designed to give maximum natural light for working. Now that we have to use computers, the ambient light needs subduing—a sad reflection on technology.

The Shadow Collection, *designed by Marcel Wanders in 1998 for Cappellini. The exaggerated scale of this floor light is clearly a comment on the traditional table light. This light, however, is made from a translucent composite film hung over a skeletal metal rod frame with incandescent bulbs both in the shade and the stem. The effect is an even, glowing, ambient light, which works particularly well against a wall or to open up a corner.*

There is clear scientific evidence that light profoundly affects the psychology, spirit, mood and emotions. This is most pronounced in territories near the poles, such as Scandinavia and Iceland, where there is very little natural daylight for much of the winter. Consequently, in these countries there is a high incidence of cyclic clinical depression.

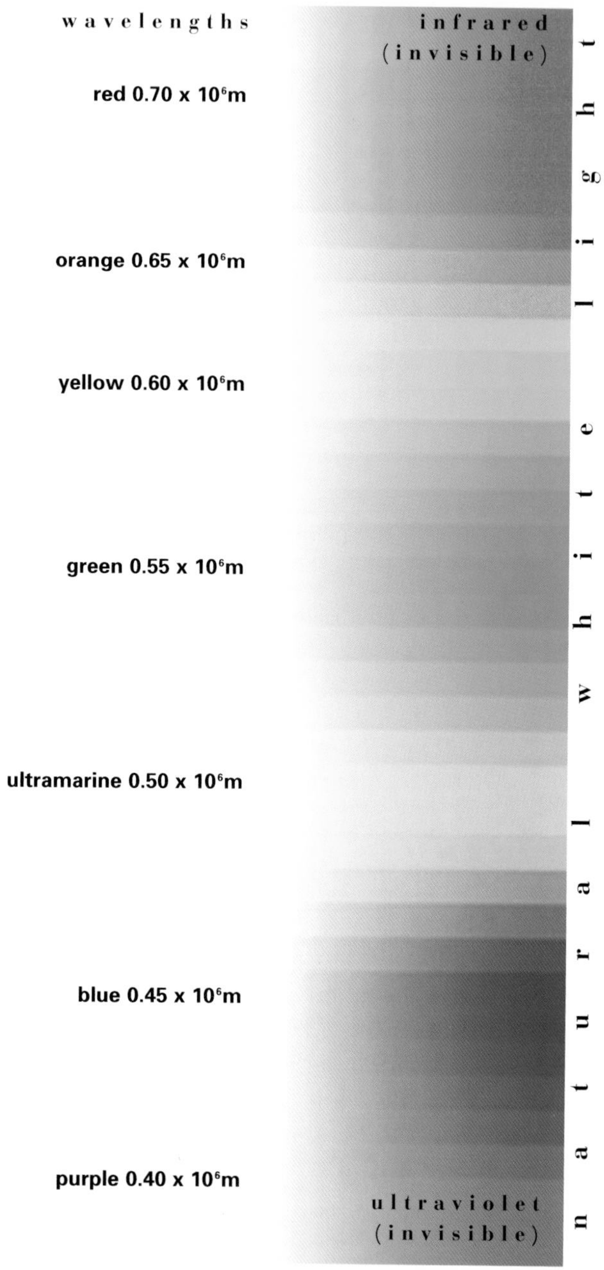

wavelengths

infrared (invisible)

red 0.70 x 10^{6}m

orange 0.65 x 10^{6}m

yellow 0.60 x 10^{6}m

green 0.55 x 10^{6}m

ultramarine 0.50 x 10^{6}m

blue 0.45 x 10^{6}m

purple 0.40 x 10^{6}m

ultraviolet (invisible)

natural white light

what we perceive as "white" light. We may take it for granted, but imagine life without it, without sight—most of what we know comes through our eyes.

Scientifically speaking, light is generally a product of intense heat exciting molecules of matter to such an extent that they rid themselves of energy by radiating it as light. Different chemical elements tend to give off individual wavelengths (their spectrographic signature), but in general, the hotter the source, the brighter the light. Therefore, all lamps give off different quantities and qualities of light with subtly different properties depending on what they are made from and how they are constructed and powered. Accordingly, they are suitable for different applications, since no lamp can give as broad a spectrum of light as that derived from the extremes of the ongoing nuclear fission of the sun.

As a general rule, the higher the temperature of a light source, the greater the spectrographic bias to the shorter (bluer) wavelengths of the light. In other words, proportionally less infrared gives a broader light that brings out the blues as well as the reds intrinsic to any object bathed in it. It does get a little more complex, as the ingredients of the spectrographic cocktail of each type of light are not equally proportional, but this is not the place to go into it.

Here is a pertinent example: candlelight has a higher proportion of warm-colored long wavelengths than moonlight, which has a bias to cooler-looking shorter wavelengths. Hence warm, rich reds and skin tones look better in candlelight than in bright moonlight, despite the fact that the two light sources are similar in intensity. Likewise, sunlight has a very broad color spectrum whereas a fluorescent tube's is narrow (this is what makes it energy efficient). An example of narrow waveband light is the yellow low-pressure sodium lights that are used on highways. These are particularly efficient in terms of output brightness, but across a very narrow spectrographic band. Color temperatures of various sources are typically:

Until the Second World War, acetylene gas was used in portable lamps on early bicycles and automobiles. Acetylene gas provided a relatively high-intensity light because it burns at a much higher temperature than coal gas, which was used at the time. Acetylene gas can also be created by dripping water on portable calcium carbide granules, as shown here where Tom Dixon is enjoying welding with acetylene gas.

Incandescent: 2700K
Quartz-halogen: 3000K
Fluorescent discharge: 3000K (warm); 3500K (neutral); 4100K (cool white); 5000K (daylight)
Evening sun: 3000–3500K
Cloudy midday: 4000K
Bright midday sun: 5000K

Objects are colored according to the wavelengths they absorb and reflect. A brilliant white object, for example, absorbs only about 8 percent of visible light, reflecting 92 percent of the light shone on it. A matte black object absorbs about 95 percent of visible light, reflecting very little. A red apple absorbs green light and reflects red, while a green apple absorbs red and reflects green.

The spectrographic mix of light clearly affects the perception of colors viewed in it. The eye reacts to different color temperatures too; the colder the light, the smaller the pupil. For example, fluorescent colors appear to glow brighter than their surroundings because they alter the short wavelength of ultraviolet light, which is invisible to the eye, when reflecting it, making it longer and therefore more apparent.

getting started

This spherical boulder of wax will burn and burn, creating flickering shadows. It acts more as a piece of minimalist sculpture for that occasional highlight rather than as a practical light source.

history

There are two predominant forms of light used in modern times: artificial light from electricity and natural light from the sun. They are about as similar as lead and gold—one makes a poor imitation of the other.

Without some sort of artificial light when darkness fell, early Homo sapiens would have had no option but to stop whatever they were doing and sleep. The night in winter would be long and cold, in summer short. The harnessing of light was what allowed ancient man to unshackle himself from the routine imposed by nature, and was perhaps the first real step in differentiating ourselves from the beasts.

Until the twentieth century, after dark the world was lit largely by fire. Although electric light had been around as a scientific curiosity since Sir Humphry Davy's arc lamp experiments in the early 1800s, it wasn't until 1879 that mass-produced electric light became attainable, with the development of the incandescent electric lamp. During that year, Thomas Edison (using screw fittings in the U.S.) and Joseph Swan (using bayonet fittings in the U.K.) simultaneously developed a reliable glowing filament, and electric lighting really took off.

The principle of this invention was pivotal to the discovery of suitable materials that would conduct electricity, producing extreme heat and light when a voltage was applied across them. The filament had to tolerate the white heat required to give good light without disintegrating. After much searching for the right material to use as a filament, legend has it that in Edison's case this was carbonized cotton from a female visitor's garment. This discovery, coupled with the advent of an electricity supply network, redefined the meaning of the word convenience. Now, a 100-watt tungsten filament lamp with more than a yard of coiled wire in it can theoretically emit the light of 120 candles.

The twentieth century was set to become the century of electric lighting. Initially, the design of these early electric lights was clearly influenced by lamp and candle stands of the time. Innovative design approaches for the new phenomenon quickly took root, though, as evidenced by the work of the artistic design movements of the era. With the Arts and Crafts, Art Nouveau and Successionist schools, it soon became apparent that

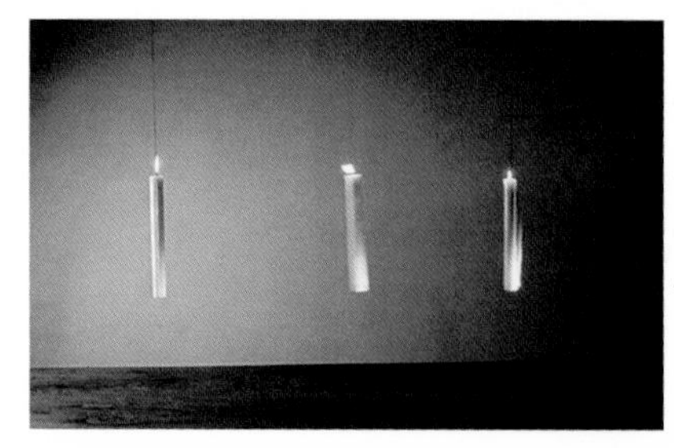

Fly Candle Fly, *designed by Georg Baldele for Ingo Maurer. These miniature pendant lights give off an almost ecclesiastical ambience from their small capsule bulbs.*

lights could be important decorative objects; early examples are the colorful stained glass lamp shades that were designed in Tiffany Studios in New York. New forms and approaches began to be developed by the other emerging European design schools that made up the Modern Movement, such as Art Deco, De Stijl in the Netherlands, Bauhaus in Germany, and the developments in Scandinavia, which were epitomized by the Danish designer Poul Henningsen (see his revolutionary 1958 *Artichoke* pendant light, opposite).

Until recently, the construction of lights for use in the home has been restricted by the technical requirements of the incandescent lamp, the constraints of the forms that are actually available and the need for efficient reflection, direction and lamp cooling. Some designers have used the lamps as a visual feature, emphasizing their forms, while others try to hide the source from view to reduce glare. The postwar period of the 1950s and 1960s saw the development of lamps with integrated silvered reflectors that biased the light to make spotlight bulbs. This allowed lights to be made with more compact heads (see *Luminator* on page 31). Taking lamps from other intended sources also became popular, such as low-voltage lamp units from cars (see Castiglioni's *Toio* light on page 45).

Electric light had been around as a scientific curiosity in the early 1800s. However, it wasn't until 1879 that mass-produced electric light became more readily attainable.

In the search to develop a smaller bulb for the automotive industry it was discovered that with the addition of a halogen gas such as bromine or iodine, the tungsten filament's vapor forms a chemical bond with the gas. This is redeposited back on the filament, allowing it to glow at 3000K, well above its normal maximum operating temperature. The result is a brighter, "cooler," more energy-efficient light that is more closely matched to natural light. The manufacture of smaller capsules for use in confined spaces is made possible by the ability of the quartz glass to resist the violent temperature fluctuations caused by the proximity of the bulb's filament to the glass.

In the latter part of the twentieth century, these quartz-halogen lamps were especially favored by the retail industry, since the absence of infrared made the light cooler and therefore particularly suitable for food applications. The development of the dichroic reflector resulted in jewel-like spot lamps that were attractive to look at, too. These truly beautiful and compact bulbs have allowed designers to make even smaller lamp-heads in their lights, which also give a better quality of light. In turn, this has created an even greater freedom in the world of lighting design as we look forward to the new century's developments.

1907
Mariano Fortuny Y Madrazo
1920
Gerrit Rietveld
1927
Eileen Gray
1932
George Carwardine
1933
Pietro Chiesa
1958
Poul Henningsen
1962
Achille & Pier
Giacomo
Castiglioni
1969
Vico Magistretti
1972
Richard Sapper

bulbs

In its present form, the "light bulb," as the electrical incandescent lamp became known, is something of a design classic, remaining more or less unaltered for almost all of the twentieth century. It has become an icon of this era, which is a statement that few industrially produced

1 200W double-ended halogen, *also available as 150–500W.* **2** 150W halogen *with Edison screw fitting.* **3** 2-pin, 50W halogen capsule *for use with a reflector.* **4** 50W open-fronted dichroic halogen miniature spot *with integrated parabolic reflector, available in various beam angles from spot to flood.* **5** 50W miniature line-voltage halogen spot *that doesn't need a transformer.* **6** Full-size flood 30–75W halogen spot *with Edison screw fitting.* **7** 21W compact fluorescent *with integrated ballast and Edison screw fitting.* **8** 2-pin, 16W fluorescent formed tube, *available in a variety of sizes.* **9** 4-pin, 22W circular fluorescent tube, *available in a variety of sizes.* **10** 18W compact fluorescent low-energy globe *with electronic ballast and Edison screw fitting.* **11** 7W compact fluorescent single tube, *available in a variety of sizes.* **12** 15W compact fluorescent double tube *with electronic ballast and Edison screw fitting.* **13** 10W pencil-thin mini fluorescent single tube, *useful for discreet undershelf lighting.* **14** 15W single tube fluorescent *available in a variety of color temperatures depending on tonal qualities required. Needs an electronic ballast fitting.* **15** 525mm compact fluorescent U-tube.

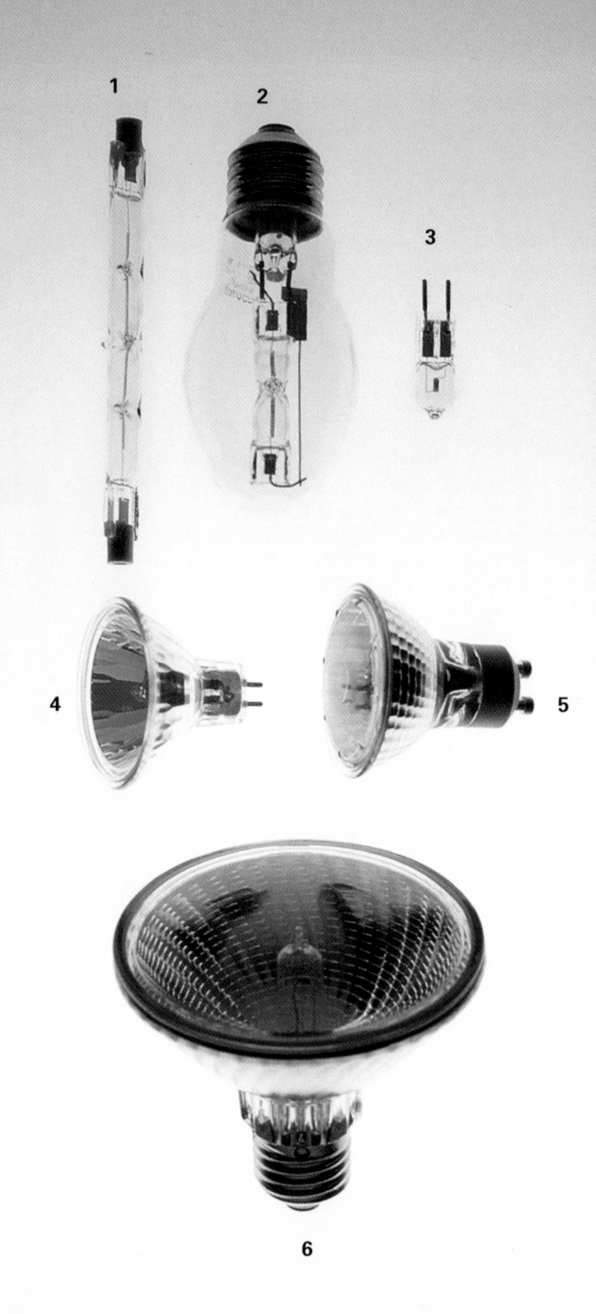

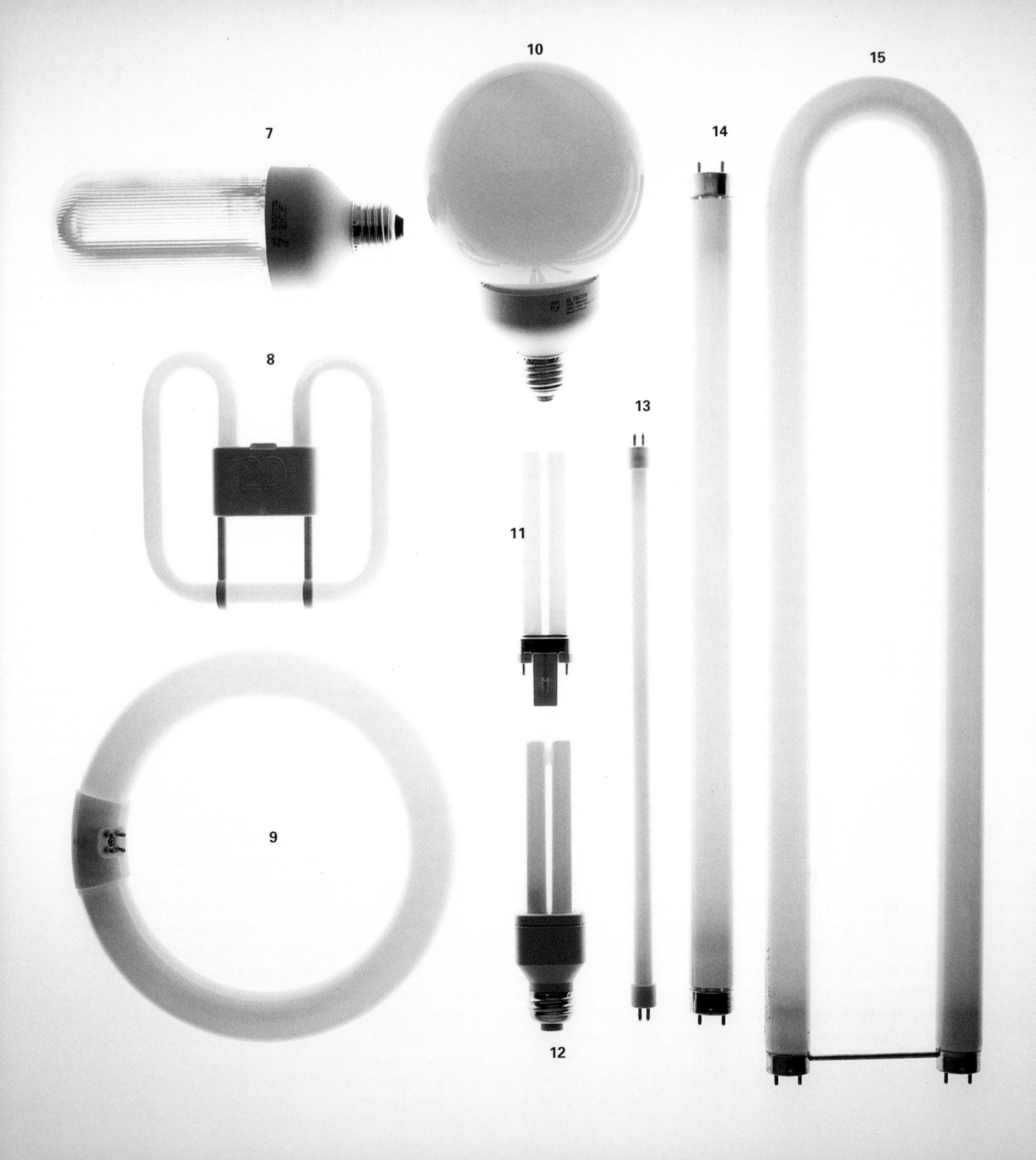
10
15
7
14
8
13
11
9
12

items can claim. Conventional incandescent bulbs (tungsten filament is the best as it lasts longer) have a maximum operating temperature of around 2700K before the coiled, yard-long filament starts to vaporize. Also, they have to be quite large; otherwise the glass would shatter under the rapid, extreme temperature changes caused by switching on and off. But over the last twenty-five years there have been many technical innovations, such as the introduction of low-voltage quartz-halogen lamps and integrated dichroic reflectors.

Recently, however, the faithful light bulb seems to have fallen from grace. Having once been the coolest high-tech gizmo of the 1880s (its image even became the universal symbol of a bright idea), the naked bulb subsequently came to represent austerity, too. Technology has not stood still, however, and there are now many new alternatives to the familiar bare light bulb.

The other ubiquitous light source is the economical-but-bleak fluorescent tube. Once lighting's equivalent of the theoretically practical but prosaic low-cost-housing tower block, due to some sensitive design and technical input it is now rather more acceptable (as is community housing). This lamp produces its light by passing high-voltage electric pulses through a metallic gas (usually mercury vapor). This stimulates a discharge of short-wavelength (ultraviolet) invisible light that in turn excites the phosphor coating on the tube, causing it to fluoresce, radiating visible light.

The positive aspects of the fluorescent tube are that they are extremely energy efficient and can be effectively used to give a good "even canvas" of ambient light that can be used in conjunction with more focused light sources. There are, however, some less desirable properties. As well as their large size, they tend to strobe slightly and the color of the light emitted has a greenish tinge, both of which can be spiritually tiring.

Other principal contenders in the domestic market are, in various forms, the compact fluorescent or the ineptly named "economy bulb" (who would want to admit to having an economy of style?). Some of the most exciting developments in fluorescent lamps are the pencil-thin miniature tubes. The future looks bright too for fiber-optic cables, since they can beam a surprisingly bright pool (35W+) of cool light from the tip of a slim cable, keeping the bulky power source in a hidden, remote location.

When choosing your light bulb, remember that the smaller the point source, the more defined and darker the shadows and the more "twinkly" the reflections. In addition, the further the source, the sharper the shadow. Conversely, a broad source of light will give more diffuse shadows and softer reflections. Clear glass filament bulbs also give a point source with sharper shadows, whereas opalescent bulbs create more diffuse shadows.

1 100W Blacklite globe bulb *with Edison screw (ES) that has a high ultraviolet output for bringing out those whites.* **2** 60W silver crown ES *that hides the light source, reflecting it back on a parabolic reflector to ensure low spillage from spots.* **3** 40W bayonet-fitting opal "golf ball" *incandescent bulb.* **4** 25W mushroom reflector *with small Edsion screw (SES).* **5** 15W miniature bulb *with bayonet fitting.* **6** 40W silver crown with SES. **7** 25W candle flame-shaped bulb with SES. **8** 60W full-size mushroom spot *with bayonet fitting and integrated reflector that suffers slightly from light spillage.* **9** *The classic light bulb—a* 60W opal *with ES.* **10** 120W spot *with ES, available in different colors. Originally developed for outdoor use, these throw a good beam with little light spillage.* **11** 60W daylight simulation bulb *with bayonet fitting. Gives a better simulation to natural daylight by filtering out much of the warmer spectrum.* **12** 60W integrated reflector spot *with bayonet fitting.* **13** 60W, 45mm clear "golf ball" *with ES.* **14** 25W flicker flame *with small bayonet fitting and oscillating effect to replicate a flickering candle.* **15** 15W ball colored miniature bulb *with bayonet fitting.* **16** 25W opal candle bulb *with SES.* **17** 100W high-output halogen mushroom reflector *with ES.* **18** 30W, 284mm, long clear striplight lamp, *used for hidden undershelf lighting.* **19** 60W bayonet fitting, *mainly used in candelabra fixtures.* **20** 35W twin cap *opal architectural lamp.*

1
2
3
4
5
6
7
8
9
10
11
12
13
14
15
16
17
18
19
20

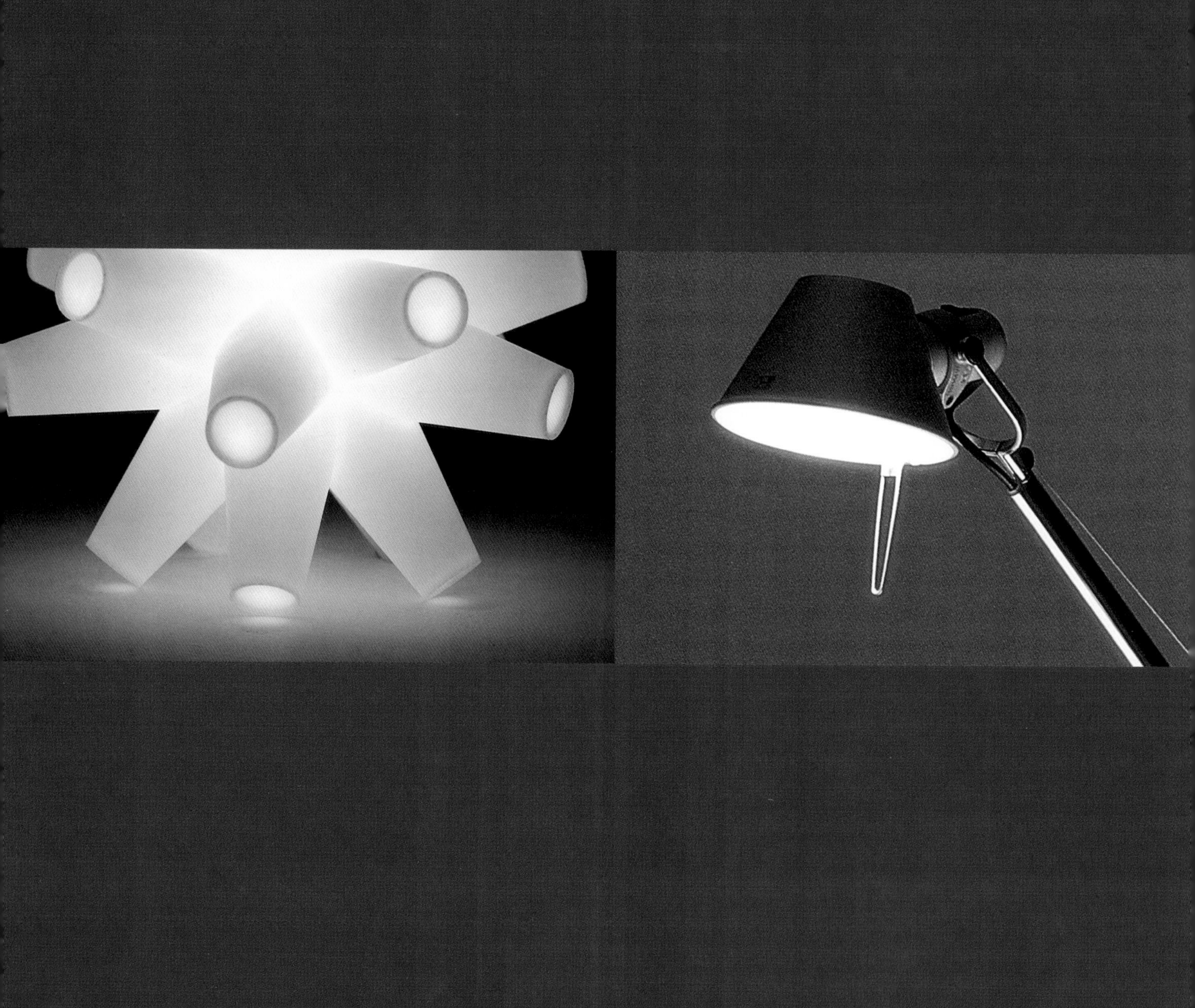

types of lighting

The creative use of light is as much about the use of shadows to give form and drama as about making things blandly visible. This can be supplemented by adding interest using color, texture and pattern. The practical side of lighting is also concerned with health and safety, particularly in the workplace and in public areas.

We may think of daylight as a constant type of light, but it does vary dramatically depending on the weather, the season, the time of day, the location (the seaside tends to have cleaner air and cooler light)—even the latitude has an effect.

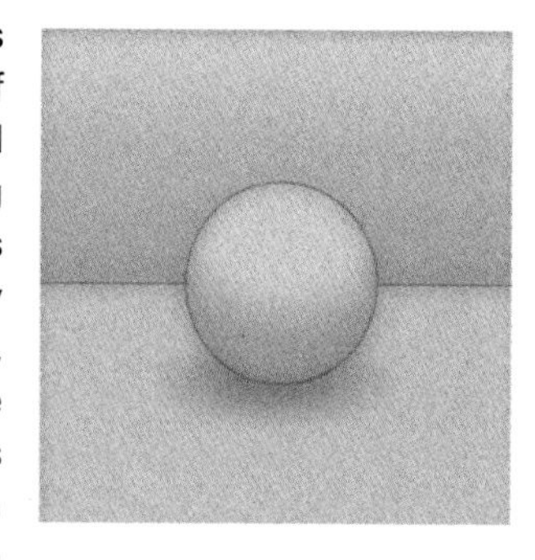

AMBIENT LIGHTING

Ambient light *is similar to the soft, even light that you get on an overcast or cloudy day, with very little or no shadow. It is easy to see by, but everything looks a little bland without form-defining shadows.*

DOWNLIGHTING

Downlight *is similar in effect to the light received at midday on a cloudless day, resulting in intense shadows. Downlights are usually used in multiples or with some ambient light to color in the dramatic shadows.*

UPLIGHTING

Uplight *is usually reflected off the walls and the ceiling and is the most flattering form of lighting, since it irons out wrinkles and gives the illusion of more space. It can also create interesting architectural effects.*

Artists and painters prefer their studios to receive northern light because it has a constant intensity and consistent spectrographic mix. Consequently, the painters' colors don't change significantly in appearance throughout the day.

The degree of artificial lighting required in a room is going to be strongly influenced by the intended use of the space, coupled with the amount of natural light available. So careful consideration must be given to what the environment will look like both during the day and at night. Our bodies have been genetically programmed to respond to light levels: an overly bright environment will make us feel alert and wakeful; conversely, subdued light makes us feel more relaxed. Too much bright light for too long makes us feel tense; too much time in subdued light is depressing to the psyche.

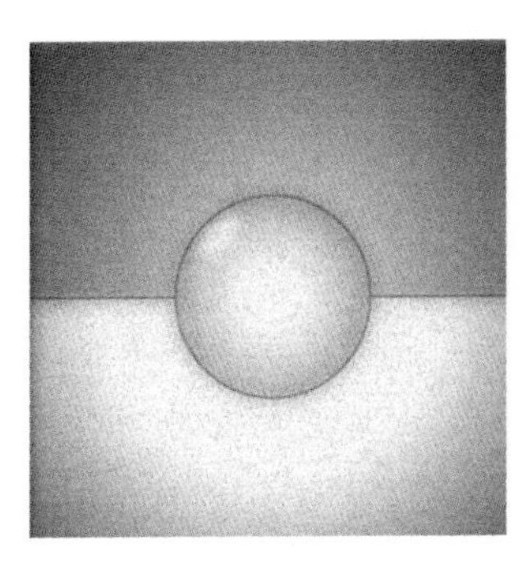

SPOT LIGHTING

Spot lighting ***can highlight objects in a dark room, create pools of light on walls or provide useful reading light over seating areas or beds. Spotlights generally need to be used with a reasonably high level of ambient light.***

CANDLE LIGHTING

Candlelight ***creates long, hard, flickering shadows, which can be both dramatic and romantic at the same time. The shadows can be softened to good effect if candles are used in multiples around a room.***

TASK LIGHTING

Task lighting ***provides a bright pool of light over the task at hand, making it easier to see and concentrate. Because of their flexibility, use task lights to good effect by reflecting the light off walls and surfaces.***

OCCASIONAL LIGHTING

Occasional lights ***tend to draw attention to themselves rather than provide useful light; they work best with subdued ambient light, when their lighting effect is most noticeable.***

Wicker Floor Light Range, designed by Michael Sodeau in 1998. Reminiscent of lobster pots, these lights are more sculptural than a source of practical light. They are particularly pleasing when used in multiples and the shadows are especially effective if the lights are used near a wall.

ambient lighting

An overemphasis on bland surrounding "ambient" light is not only wasteful but also a little soulless, since rooms and objects without defining shadows lack form and character. Conversely, a room full of impenetrable shadows may look dramatic but is uncomfortable to live or work in. Instead, some ambient light is pretty much essential for filling out the shadows.

Often, you can best achieve ambient lighting by reflecting light off the walls (by using ceiling- or floor-mounted wall washers) or off the ceiling. An alternative approach for good ambient lighting is to use multiple sources, such as standard lamps that are casually dotted around the room with large diffusing shades. Go ahead—create light pools where they are needed. Although all rules can be broken, it is generally a good idea not to make the primary light source (the lamp itself) visible. Instead, always have the bulb diffused or reflected by the shade rather than allowing it to spill its glare and attendant harsh shadow.

With the proliferation of computers the requirements for work lighting have recently changed. Instead of a pool of bright light on the work surface, subdued, diffused, ambient light minimizing overspill from primary light sources is preferable to prevent glare or reflections on the display screens.

During the course of a day, a kitchen/dining room, for example, is used for preparing and eating breakfast, reading the paper, maybe some domestic administration, coffee and a chat, preparing supper, enjoying the evening meal and perhaps indulging in a little television, too. In the morning, the space may be bathed in sunlight from the east and in the evening, have no natural light at all. Flexible lighting is required, not just an on/off switch.

Ambient and some task lighting is required for preparing food, but supper will be much more enjoyable—in fact, it will actually taste better—if it is not eaten in the same bright, diffused light. One approach is to have different combinations of lights that you can vary to change the mood. Dimmers are useful for this but check to see that they are compatible with the light source. You will find that fluorescent lamps and some low-voltage transformers may not perform with dimmers or even may cause them permanent damage.

downlighting

The light source for downlighting can be ceiling mounted or recessed, or it can emit from ceiling spots, edge and corner bulbs, pendants and wall-mounted floods. There is much that lighting can do to change the perceived size and proportions of a space. The most noticeable effect of downlighting is to seemingly drop the height of a ceiling. The ceiling is cloaked in shadow and therefore appears to come closer to the viewer.

Always bear in mind the direction of light in relation to the position of the eye. Look at a photograph that was taken in full sunlight and then look at the same subject photographed with the sun behind the camera. The difference could not be more pronounced—in the first the subject is obscured in shadow and in the second it is well lit with all features visible. The position of light is paramount to our perception of the "subject" that is illuminated. Any half-decent baseball player, fighter pilot or interrogator will tell you of the advantage of having the light behind your shoulder. Although dramatic and flattering to architecture, light polarized from directly above puts the eyes into shadow, giving a hard impression. Having said this, downlights are useful for enlivening dull corridors or hallways. When used in evenly spaced multiples with a broad spread of beam, downlighting offers a good level of diffused lighting that is similar in shadow effect to light clouds at midday. For a more dramatic effect with sharp, dark shadows similar to those of a spotlight, try using a single, slim downlight beam. General downlighting in a bathroom is not a good idea as it creates shadows, making it difficult to shave or to apply makeup. A downlight over the toilet area, however, is useful if you are partial to reading on the throne; and a downlight directly over a bath can make you feel as if you are bathing in light as well as water.

If the intention is to use the space for work or entertainment, ensure that there is enough ambient light to fill the shadows. If you have the room height, pendant downlighters work well at reinforcing ambient light by creating pools of light over dining tables or work-surfaces as well as providing a focus to the room. The crystal chandelier is an extreme example of type.

The **PH Artichoke Copper**, *designed by Poul Henningsen in 1958 for Louis Poulsen, is a pendant fixture, housing petal-like segments. Suspended from three steel wires, this light is cleverly designed to give an even diffused light while concealing the light source.*

uplighting

Uplighting comes in many varieties and flavors, varying from wall-mounted sconces and adjustable-angle floodlights to floor-level wall washers, table lamps and high-level standard lights. Low-level uplights tend to be especially effective when used in reception rooms and other areas. The absence of glare from the lamp is an important aspect of the received impression.

The fundamental principle of uplighting is the reflection of light off the walls or ceiling and into the room. This gives good, indirect diffused lighting, defining the parameters of the space with more interest than ceiling-mounted direct lighting. The really magical effect of uplighting is that it appears to lift the ceiling height by making the ceiling appear lighter and therefore further away. As such, it is quite perfect for low-ceilinged rooms and in spaces that you would like to appear larger.

The effect uplighters give is affected strongly by the color and type of wall covering that they are directed toward. Stripy wallpaper with vertical lines leads the eye upward, creating the appearance of added space. In a narrow room, horizontal stripes widen the space by leading the eye along the wall. Combining lighting with these visual tricks adds to the illusion of more space. When lighting washes over walls, it is important that they are well finished or flat, since the light highlights the tiniest of imperfections with oblique shadows. Conversely, this can be put to good use on textured walls, such as exposed brickwork, where it will accentuate the character of the surface.

Low-level uplighting also provides any room with warmth. Furniture lit from behind with uplighting can look dramatic while giving the impression of more space. Certain table lamps can be effective in creating uplighting in the middle of a room while offering an interesting sculptural accent to the environment.

Above *The* Jill Wall, *designed by King, Miranda and Arnaldi in 1978 for Flos. Here is a classic use of a double-ended halogen flood bulb housed so that most of the light is directed up while the cast glass fitting glows in the spillage. The design is also available as a floor-standing uplighter and the glass shade comes in yellow, pink, white, green and medicine-bottle blue.*

Opposite *(background)* **and below** Luminator, *designed by Achille and Pier Giacomo Castiglioni in 1954 for Flos. This floor lamp uplighter was one of the first fixtures designed around the integrated reflector mushroom spot lamp.*

Opposite *(foreground)* Atollo, *designed by Vico Magistretti in 1977 for O Luce and first made of lacquered aluminum. Since then it has been made from acrylic (1988) and Murano glass (1989). A geometric take on the classic table lamp. Atollo directs all light downward onto the horizontal surface below.*

spot lighting

Spotlights derive from automotive headlights and the full-scale versions that are used in theaters to cast a small pool of light (or "follow" spot) on a particular performer or action on the stage. They have a distinct advantage over other forms of directional lighting in that they tend to be quite small for their output, so they can be easily mounted on ceilings or walls, and often they can be set on freestanding pylons too. They generally work by using a parabolic reflector behind a light source to produce a parallel beam of light that does not diffuse over distance. Sometimes this is incorporated into the bulb itself. Lasers are the most extreme form of spotlight, with a parallel beam angle of zero degrees.

The average domestic spotlights are not nearly so narrowly focused; their main use is to project a well-defined, bright pool of light highlighting a particular area or object. These vary in intensity and geometry, but generally they tend to be quite small and have a fairly narrow angle of beam compared to floodlights.

The consequential effect of spotlights is that their sharply defined, dark shadows can be very dramatic, bleaching out any color if there is not enough ambient light too: the eye assimilates to the darkness rather than the light. This makes them particularly good for highlighting objets d'art, sculptures or pictures at a distance from a ceiling or across a room. Spotlighting a picture from recessed lighting creates the illusion of more space. An entry hall or corridor is a good place for display; try spotlighting objects and art here to create an intimation of the style of the rest of the home.

The kitchen is a treasure trove of interesting objects that can be highlighted and picked out with single spotlights. Mundane pots and pans can become jewel-like decorative objects with selective lighting. Spots can also be used to shoot pools of light onto central work surfaces, and in the bathroom, you can shoot spots onto baths and basins to see light dance on water. They can be put to good use as bedside lighting, too, since they can be aimed at pillows and are indispensable for reading in bed while not disturbing your partner.

Papiro, *designed by Sergio Calatroni in 1988 for Pallucco Italia. With a flexible and directable adjustable floor lamp, the base, stem and diffuser are made from copper, with a nickel-chrome finish. Available in two heights, the lights come with an electronic transformer and dimmer with foot control lever. This creates a pool of light by moving the light source close to the subject rather than projecting a narrow beam from a distance.*

Nondrip paraffin wax candles can be an inexpensive and effective source of mood lighting if used in multiples, as here, making a feature of this disused fireplace. Be careful to never leave candles unattended.

candle lighting

These days we take it for granted that when we enter a dark room we can flick a switch and . . . presto! we have light. But it was not always so. Our earliest ancestors probably first saw fire when lightning struck a tree. Soon man learned how to create, nurture and control fire himself, making torches, oil lamps and candles for convenient light sources after darkness—extending the useful part of the day.

Today's decorative, festive candles have their ancestry in what was once the principal source of artificial light—mutton-fat tallow candles for the poor and beeswax for the rich and the church. We now use paraffin wax derived from crude oil. Then, with the advent of coal gas in the early nineteenth century, efficient lamps were developed with the invention of the gas mantle, giving a far brighter, whiter light for both street and home.

Candles, however, have always been far more elegant and stylish than plain and economical oil lamps, especially when used in multiples. Today we rely on them more for atmosphere than for light. Characteristically, candles tend to give a very warm light with long, well-defined flickering shadows caused by their small light source. This is coupled with the soft glow of the wax, illuminated by the hungry flame that is soon to consume it.

The warm glowing light of a candle is famously flattering if used in the center of a table, since it hides wrinkles and imperfections. When they are used with abandon they give wonderful character to a room in the evening.

Votive candles have the advantage of being inexpensive and lasting for a good eight hours on average. This makes them particularly useful, since they can be lit early without the need to be serviced. They do need some help with their appearance, though. Naked they can look a little basic, but there are many specific lamp holders for them and they also give ample opportunities for improvisation.

Think safety—always take care with fire. Think what will happen when candles burn low, be careful of nearby flammable objects and don't use wooden holders. Never leave a room with candles alight—especially to go to bed.

task lighting

Task lighting is created by floor or table lamps with cantilevered arms that can be easily adjusted so that the lamp throws a bright pool of light onto a zone of concentration for, say, reading or drawing. Typically, this light will be about five times the intensity required for the rest of the room.

Surgical operating rooms are painted white for good reason. Not only does it aid hygiene, but the white also reflects and optimizes a bright, even light throughout the space. There is also the familiar multiheaded (to avoid shadows) cluster of small spotlights suspended over the job at hand, creating an intensely bright pool of light on the area. This is an extreme but pertinent example of task lighting in action: the brighter the light, the smaller the pupils, so the more clearly the surgeon can see. The importance of good task lighting is often misunderstood or overlooked, especially in the home.

Task lighting's raison d'être comes down to the physics of biology. When relaxed, the eye is meant to focus in the middle distance. When looking around at objects near and far the eye has to work continually to adjust its focus to see clearly—the closer to the object, the harder the lens muscles have to work. Typically, when reading a newspaper, the focus can constantly be ranging from 12 to 24 inches, which can be quite strenuous to the eye, especially in poor light. The brighter the light intensity on the page, the more accurately the iris can be adjusted to let in the right level of light required to stimulate the retina. This, in turn, allows a wider depth of focus, thereby reducing the amount of eye fatigue from all the muscle activity needed to see clearly. Simply put—the more light, the less tiring: hence the value of task lighting.

1 Archimoon Tech, *designed by Philippe Starck in 1998 for Flos. Used with a 35W low-voltage halogen capsule, his light has a similar articulated balance mechanism as the seminal* Anglepoise *by George Carwardine (see page 17). The shade is available in various translucent colors.* **2** Tizio, *designed by Richard Sapper in 1972 for Artemide. Here is a table lamp with double luminous intensity and revolving arms and head. It is made from painted metal and thermoplastic resin and the power supply is in the cylindrical base. The two cantilevered arms ingeniously conduct the 12V DC current to the capsule bulb. Tizio is also available with a useful floor stand.* **3** Archimoon Classic, *also designed by Philippe Starck in 1998 for Flos, is similar to the Archimoon Tech shown alongside it. This light uses a conventional incandescent 60W bulb.* **4** Tolomeo, *designed by de Lucchi and Fassina in 1987 for Flos, has arms and a head that revolve in all directions. Made from polished and anodized aluminum, this utility-styled light system comes in many different forms, including articulated wall-mounted, floor-standing and with clamp heads.*

4
3
2
1

occasional lighting

There are certain lights that have such a conceptually extraordinary appearance that they cannot be easily categorized. Often weird and alien-looking, they can seem like the domestic equivalent of miniature UFOs. If furniture is to be likened to clothing for the home, then some lights are surely the accessories and jewelry. Lights can be seen as functioning pieces of mechanical sculpture that add a highlight to any environment. The light-giver becomes a working art object, and because lights are inherently bright, they attract attention to their physical forms as emitters of radiant energy. After the chair, the light is probably the nearest example of the designer's equivalent of the artist's self-portrait. Do not confuse these designer-ego-expression lights with the more functional architectural fixtures whose aim is to provide usable light to a space. Oh no...the purpose here is to delight and stimulate—providing focus and interest to an environment.

Often very strong and dominating, these lights become modern treasures with a history and distinct personality that you can love and carry with you throughout your life.

These are not really lights as such, they are more ideas that happen to emit light. There has probably been more demonstration of conceptual ingenuity in the field of lighting than in any other area of environment-related design. It is the eclectic mix of form with technological function that makes these lights the architectural equivalent of richly glittering jewels.

It is difficult to advise on how to use these objects as lighting since their output is so varied in both quantity and quality. It is their actual appearance rather than their light to which they owe their existence; therefore they may be suitable for adding interest to a dark corner or character to an otherwise lifeless room. Such occasional lights can also be a very useful means of adding color to a monochromatic environment.

Sasso, *designed by Caterina Fadda in 1998. More sculptural poetry than a conventional light source, the red glass "pebble" only lights up when placed in contact with the metal strips on its Perspex base.*

material choices

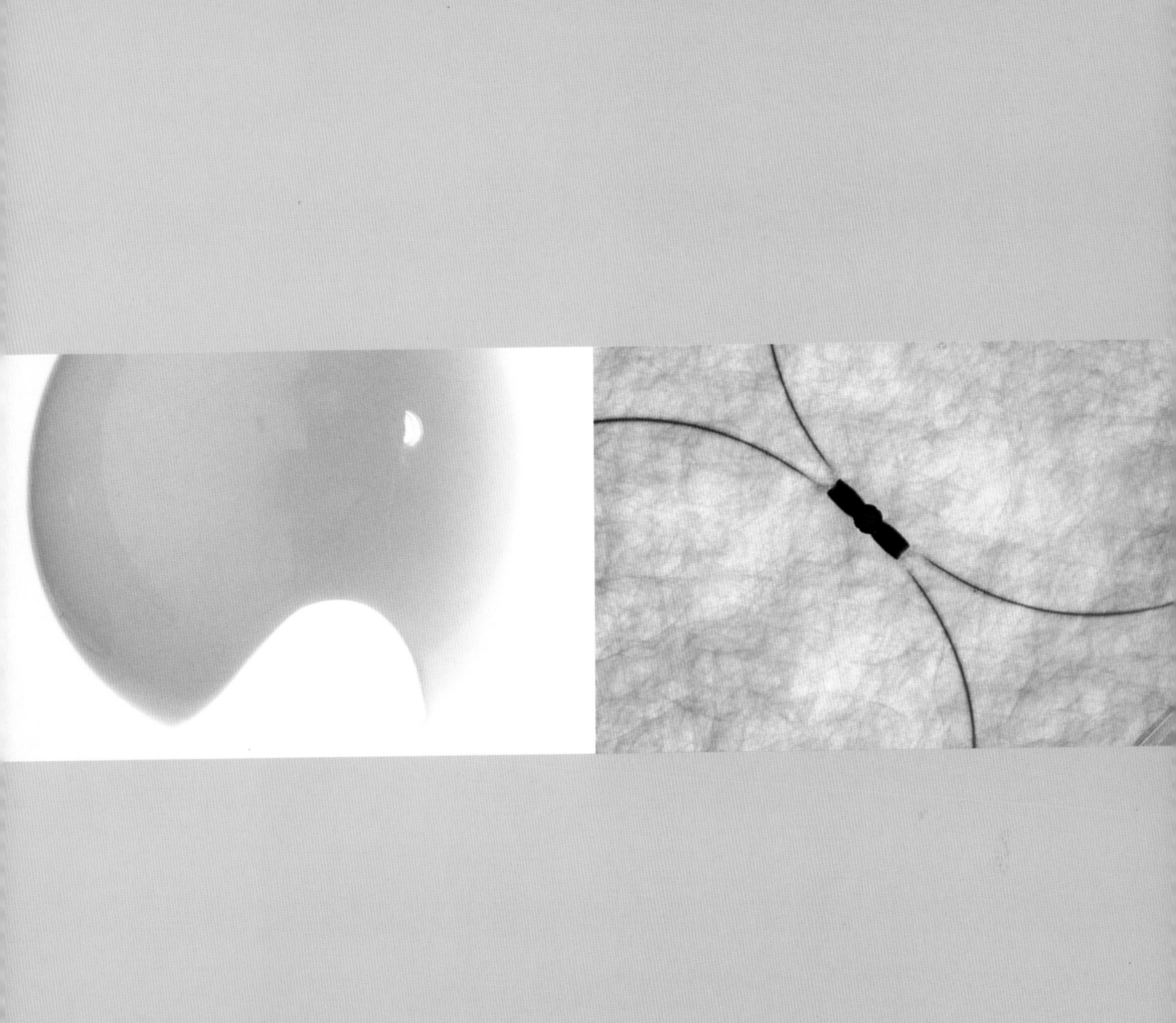

lights

It is important to have the right type of light for the right situation. A light suitable for a desk is different from that for a living room. Sometimes rooms also change in the way they are used throughout the day, in which case the lighting arrangement must be flexible. Aim to achieve a balance between ambient light and glowing pools, perhaps with accents of tightly focused point sources.

1 Taccia, *designed by Achille and Pier Giacomo Castiglioni in 1962 for Flos, gives an even light by reflecting the light off enameled metal.* **2** Costanza, *designed by Paolo Rizzatto in 1986 for Luceplan (see also page 52).* **3** Pod Lens, *designed by Ross Lovegrove in 1998 for Luceplan. Made of injection-molded polycarbonate, this is primarily an outdoor light, since the material is resistant to ultraviolet rays, humidity, water and snow.* **4** Romeo Moon, *designed by Philippe Starck in 1998 for Flos. Available with glass or fabric shades, this table light is part of a line that offers coordinated lights for pendant, wall and floor.* **5** *The pendant light from the* Terra Lighting Range *designed by Pedro Silva Dias in 1997 for Proto Design. This small, ceramic pendant lamp is useful for providing accent lighting.* **6** Espiga Lamps, *designed by Sharon Bowles and Edgard Linares in 1998. Here are occasional floor lights designed to give some interesting shadow effects.* **7** Elvis, *designed by Charles Williams in 1998 for Fontana Arte. The innovative way the crumpled translucent fabric is cast into the clear resin of the shade gives this angular table light an intriguing elegance.*

4
^
5
6
7
v

floor lights

Living rooms are the natural habitat of the floor lamp. The light in these surroundings, which are used primarily for social interaction, should make people in them look as good as possible to each other. To create a warm atmosphere, it's best to try using a variety of different light sources at varying heights (remember, low is flattering). To give a greater feeling of intimacy, emphasize pools of light around seating rather than having an even, bland light. Floor lamps are a good source of ambient light as is uplighting, especially if used at the side of the room, to reflect light off the walls.

Although they do emit useful light, bulky standard floor lamps placed in the middle of a room can cause clutter and take up space, so it's often a good idea to place them in corners, close to a wall, or at least out of walkways. Be wary of those with legs that stick out as well as those with topple-prone small bases (especially if you expect to have young children gallivanting about). Try to plan your lighting scheme to avoid the need for trailing cords and cables, and have floor sockets installed if you wish to use lighting in the center of a room—you can never have too many outlets, it seems.

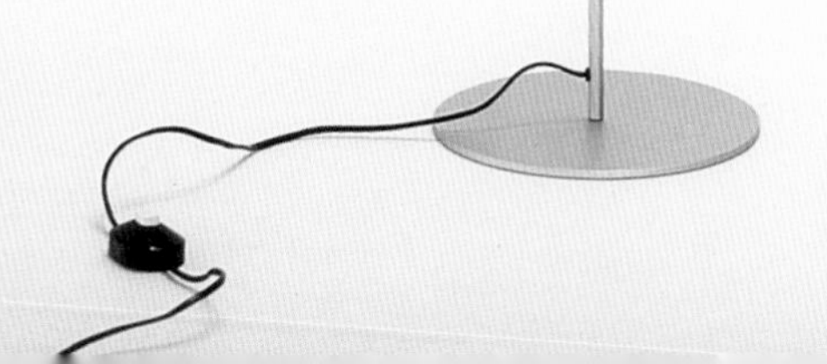

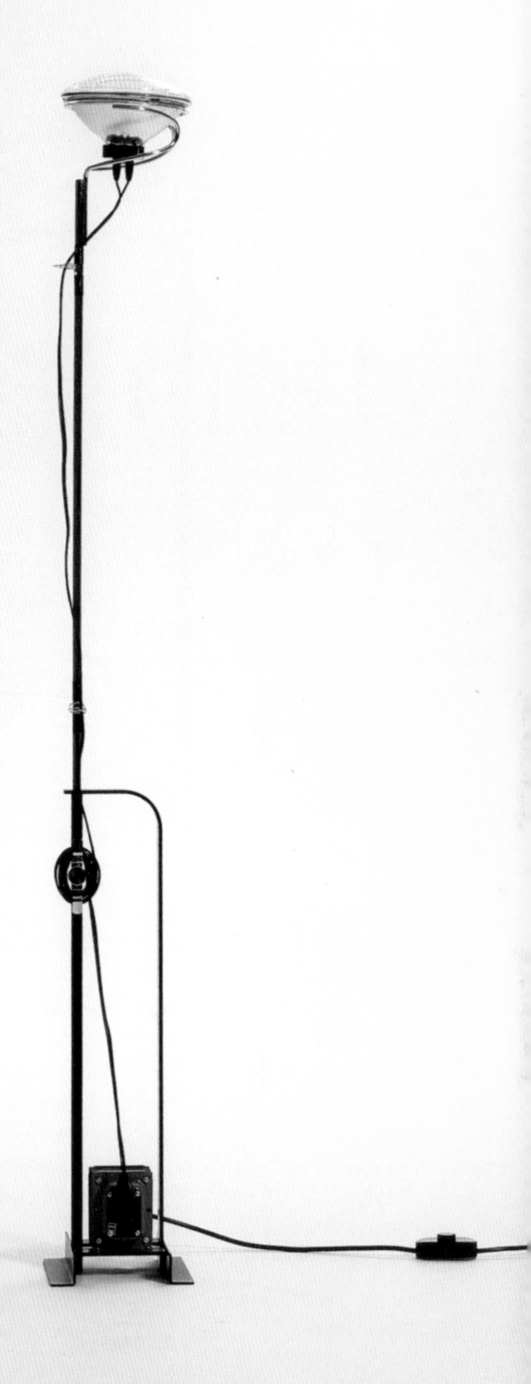

Far left Prima Signora, *designed by Daniella Puppa in 1992 for Fontana Arte. This tall, 5½-foot-high floor lamp has a large, spherical glass diffusion shade that spreads an even, ambient light from its 250W halogen light source. Elegant and imposing, this floor lamp is probably best situated near a wall or in a corner, but not where it could be knocked over by children or pets.*

Center Eclipse 3, *designed by Peter Wylly in 1996 for Babylon Design. The lamp shade of this floor lamp is made from biodegradable polystyrene sheets cut and formed into three layers. The support and base are created from mild steel. This futuristic piece is less likely to suffer damage, since it is not so top-heavy and has a lightweight plastic diffuser.*

Right Toio, *designed by Achille and Pier Giacomo Castiglioni in 1962 for Flos. This height-adjustable floor lamp gives out direct light; the fixture is set on an enameled steel base and nickel-plated brass stem. This classic light was one of the first domestic lights to use low-voltage tungsten-halogen lamps as developed for the automotive industry.*

1 Gilda, *designed by Enrico Franzolini in 1997 for Pallucco Italia, has a large lamp shade made from synthetic parchment paper.*
2 Lowlight and Highlight Lamps, *designed by Helene Tiedemann in 1996 for David Design. These cute lamps have a quirky, animal-like character that is more witty than jokey.* **3** Iride, *designed by Pierluigi Nicolin for Artemide. Remote-controlled, each light gives off a different color that can be varied according to mood.* **4** Lola, *designed by Alberto Meda and Paolo Rizzatto in 1987 for Luceplan. With its swiveling head, micro-perforated metal reflector, telescopic stem in carbon fiber and die-cast articulated tripod, this sci-fi-looking piece uses race-car technology with appropriate wishbone styling.* **5** Helice, *designed by Marc Newson in 1993 for Flos. This floor-standing lamp with halogen flood makes reference to the UFOs of 1950s science fiction. Subtle and successful colored lighting effects are obtained using glass filters.* **6** Lucilla, *designed by Paolo Rizzatto in 1994 for Luceplan, is available in a small table version, and also as a pendant lamp. The hanging shade is made of a non-combustible fabric that is similar to that used in suits worn by astronauts.*
7 *This chandelierlike construction produced by Artemide has colored glass organic shapes suspended from a metal hoop.*

2

3

1

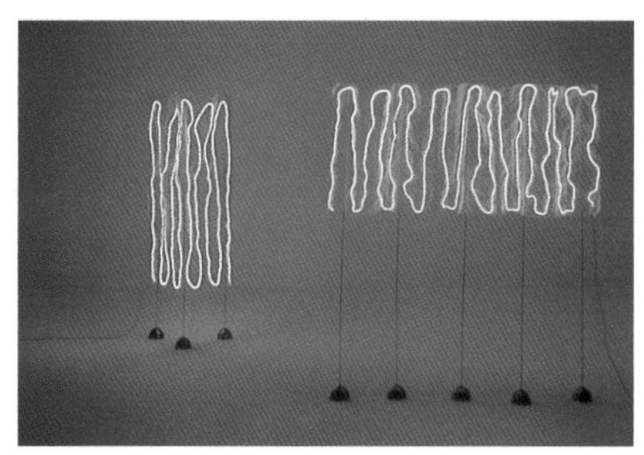

1 Long Wave, Short Wave, *designed by Bianchini Graelik Rozenberg in 1997 for Pallucco Italia, is a malleable, freestanding art light with a vibrant green-blue tone optic fiber that emits a soft light.* **2 Globlow Floor Lamp 01** *(shown inflated and deflated), designed by Vesa Hinkola, Markus Nevalainen and Rane Vaskivuori for Valvomo Design. When switched on, this ingenious floor lamp inflates via a motorized fan and continues to gently cycle between its inflated and deflated states—the effect is surprisingly serene.*
3 Papiro, *designed by Sergio Calatroni in 1988 for Pallucco Italia. This is an adjustable floor lamp with a base, stem and diffuser made in copper; it is available in natural copper color and nickel-chrome finish and in two sizes.*

Low lighting reflected off walls tends to flatter and give warmth to a space. In addition, small low and glowing lights can create useful localized pools of color that do not spill out into the rest of the space, and add interest day or night. Low-level uplighters can be used to good effect positioned behind sofas and other furniture, creating a greater sense of space. Lights that give off strongly patterned light particularly need to be placed close to walls so that their play of shadows is not lost.

Living rooms are ideal places for featuring floor lights as interesting sculptural objects in themselves, especially those lights whose output is fairly vestigial. These light-emitting, sometimes alien-looking objects can make quite a dramatic focal point in a room. Another idea is to occasionally move lights around within the home, since this can give a sense of change without affecting the ways in which the rooms are used.

Jack Light, *designed by Tom Dixon in 1996 for Eurolounge. Inspired by the children's game of jacks, these versatile light-emitting modules can be used in a variety of ways, from stools to table bases, and they can also be stacked vertically. They are made from rotationally molded polyethylene and come in a range of colors.*

Clockwise from bottom left Havana, *designed by Jozeph Forakis in 1993 for Foscarini, features four plastic diffusers suspended from each other by small metal rings.* Orbital Terra, *designed by Ferruccio Laviani in 1992 for Foscarini. The light from this collection of amorphous colored glass shapes, which are arranged up a steel spine frame, is emitted from 40W "golf ball" lamps. A less flamboyant version is also available in plain white opal glass and individually as wall-mounted lights.* Luminator, *designed by Achille and Pier Giacomo Castiglioni in 1954 for Flos, as shown in its full glory* *(see page 30).* Tube, *designed by Christian Deuber in 1997 for Pallucco Italia. Here is a fluorescent-tube floor lamp with an acrylic diffuser and metal support cradle base. It can be moved away from the base unit for a distance of up to 13 feet and may be leaned against another piece of furniture or a wall, or just laid on the floor.* 2198, *first produced by Fontana Arte in 1954, is a crisply styled piece of purist design using natural cherry with turned aluminum fittings topped by an opalescent glass shade.*

Spiral Light, *designed by Tom Dixon for Eurolounge in 1992. This tapering helix of gold- or silver-leafed mild steel with integrated dichroic uplighter is a simple and intriguing form that casts dramatic shadows. It looks almost as good when it is switched off as on.*

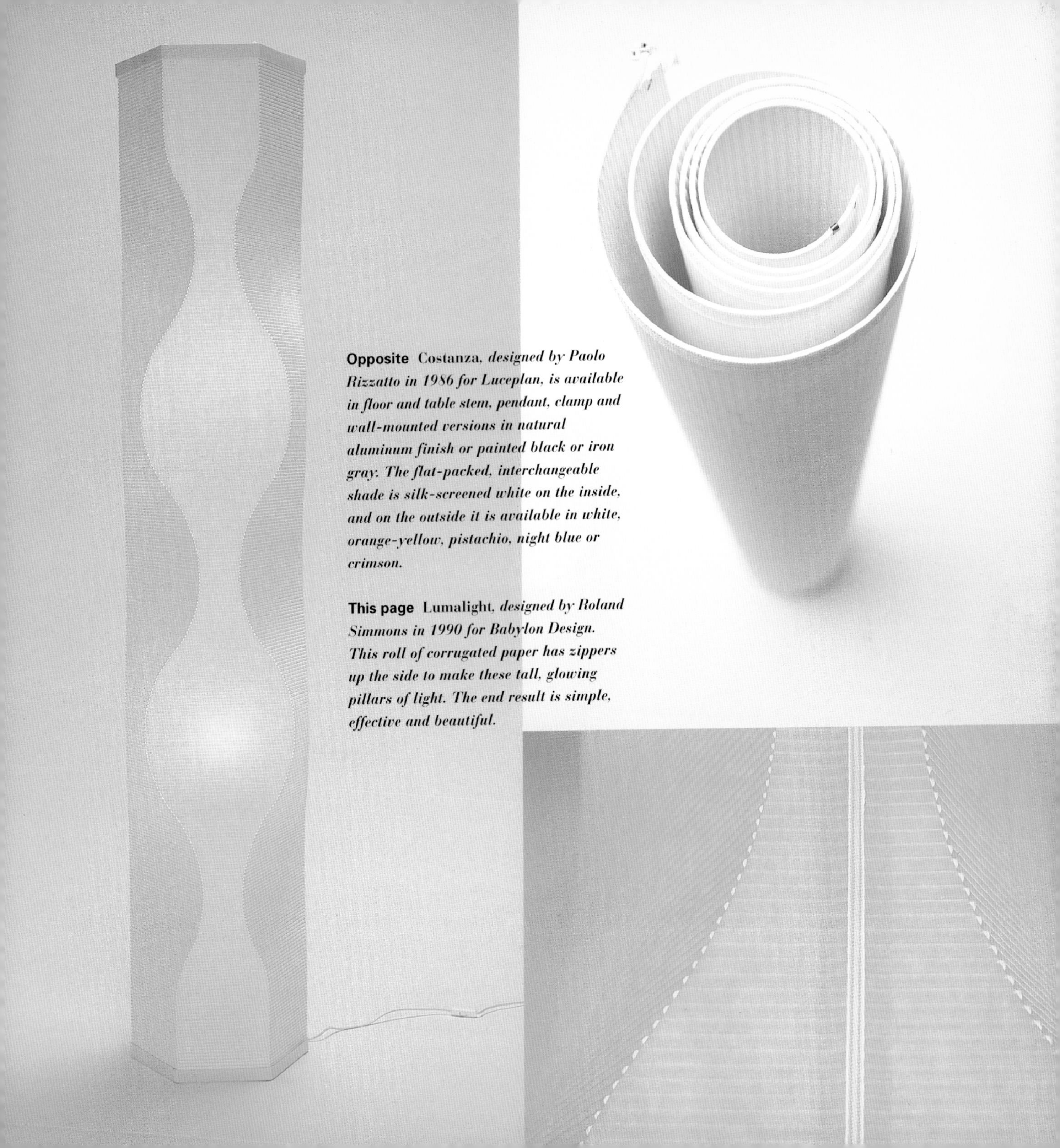

Opposite Costanza, *designed by Paolo Rizzatto in 1986 for Luceplan, is available in floor and table stem, pendant, clamp and wall-mounted versions in natural aluminum finish or painted black or iron gray. The flat-packed, interchangeable shade is silk-screened white on the inside, and on the outside it is available in white, orange-yellow, pistachio, night blue or crimson.*

This page Lumalight, *designed by Roland Simmons in 1990 for Babylon Design. This roll of corrugated paper has zippers up the side to make these tall, glowing pillars of light. The end result is simple, effective and beautiful.*

table lamps

This page, below left Glo-Ball, *designed by Jasper Morrison in 1998 for Flos. Shaped a little like a tangerine, here is a super-minimalist creation from an international design supernova.*

This page, below right Filo Table *lamp, designed by Peter Christian in 1994 for Aktiva. The swiveling polycarbonate shade directs light and is available in several colors: white, orange, green and purple. The metal base and fittings are silver-powder coated.*

Opposite page, left to right 2198 TA, *first produced by Fontana Arte in 1954 (see also page 50, bottom right).* Ilos, *designed by Pearson Lloyd in 1998 for ClassiCon. Available in three sizes, the light has a blown glass shade with ground ribbed-glass diffuser below.* Miss Sissi, *designed by Philippe Starck in 1991 for Flos. This miniature table lamp is made in brightly colored, beautifully molded plastic. It is both inexpensive and useful, giving direct light up and down and some colored diffused light to the side.* On Off, *designed by A. Meda, F. Raggi and D. Santachiara in 1988 for Luceplan. On Off is turned on and off by altering the position of its balance. The thermoplastic polyurethane shade makes this glowing mouse cool enough to handle.*

It helps to think of some lights as large vases with bunches of flowers that can be scattered around the room, radiating warm light and bonhomie, or you might view them as witty little tabletop sculptural statements. Designer-object-lights abound and are therefore what immediately spring to mind in this category. Subtle humor is often a key component in these designs, giving them character and humanity—but remember the dictum "witty not jokey, for a joke is only ever funny once." A plastic dinosaur with a lightbulb protruding from its mouth may amuse for a day or two; but how long will it be before it illuminates the closet under the stairs?

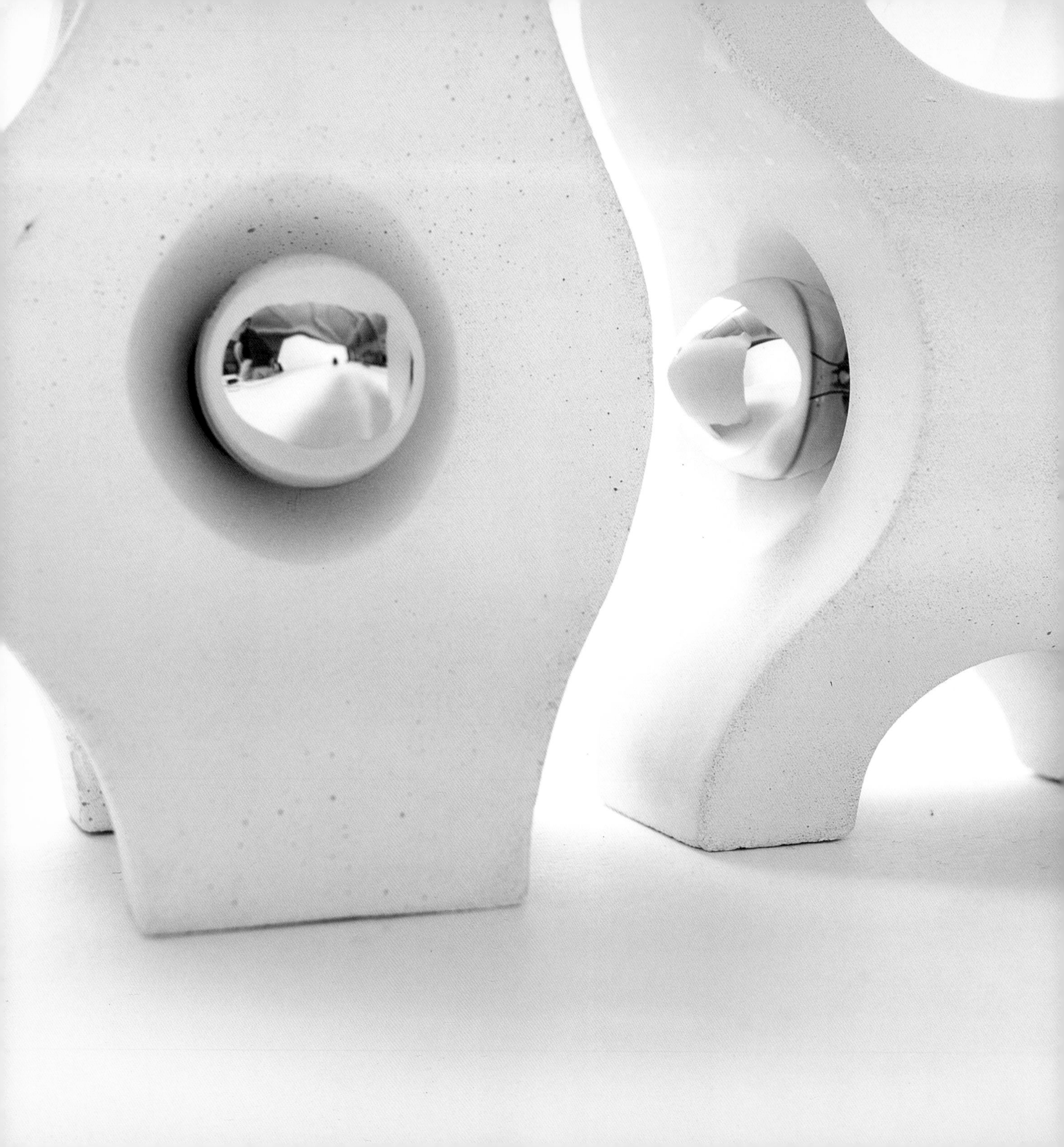

There has been a recent, unexpected interest in lights using slip-cast ceramic technology intrinsic to the manufacture of toilet bowls. They have become popular with some of the more intrepid, intuitive, independent designers, but how useful they are and how long they will be in vogue remains to be seen. They are, however, quite beautiful, having a sort of "found art" quality to them that is reminiscent of the work of Marcel Duchamp. This reference to the famous Dada art movement of the 1920s is quite deliberate and is evident in the work of many great luminary designers, including Achille Castiglioni (see pages 112–13) and Ingo Maurer (see pages 114–15).

Traditional takes on the short-stem standard lamp with semi-opaque shade continue to work successfully. The shades provide a useful control of light (especially when used with dimmers), obscuring the primary light

Left and above Oxo Light, *designed by Peter Wylly in 1997 for Babylon Design. When stacked, these cement forms bear a resemblance to the shape of the Lumalight column on page 53.*

Top right *Table lamp from the* Terra Lighting Range *designed by Konstantin Grcic in 1997 for Proto Design. This ceramic form is apparently inspired by a traditional style of candleholder.*

Right Lightbowl 1,2, *designed by Sophie Chandler for Alternative Light. This floor/table lamp uses the aesthetic qualities of classic light bulbs resting in a glass bowl, with only one switched on, to create an intriguing, glowing sculpture.*

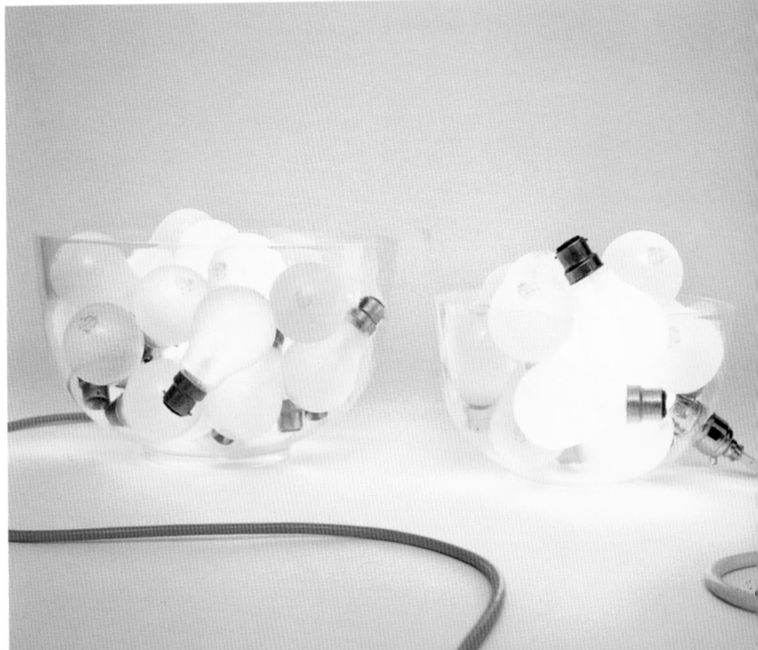

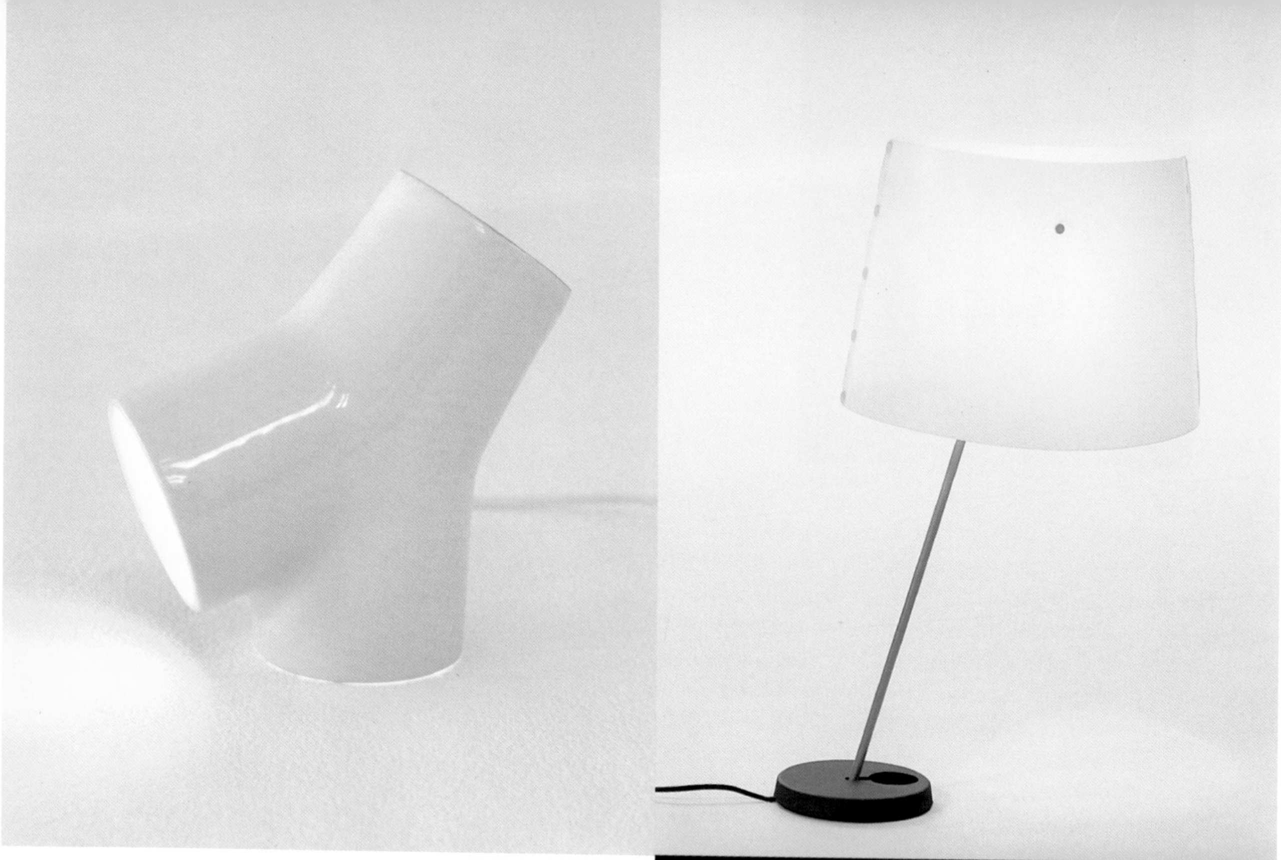

Above and opposite T13 *from the* Terra Lighting Range *designed by Sebastian Bergne in 1997 for Proto Design. This slip-cast ceramic form can be used in several ways, ranging from a table lamp to a pendant fixture.*

Top right Twist & Light 1, *designed by Sebastian Bergne in 1996 for Driade, is a table lamp with tilting, aluminum gray stem, white polystyrene shade and cast iron base, painted dark gray. The light is also available as a floor lamp with a longer stem.*

Right *The heat from this lamp, designed by Shiu-Kay Kan, rises and passes through small turbine blades, causing the patterned inner filter to rotate. Amusing for children, but it is quite delicate.*

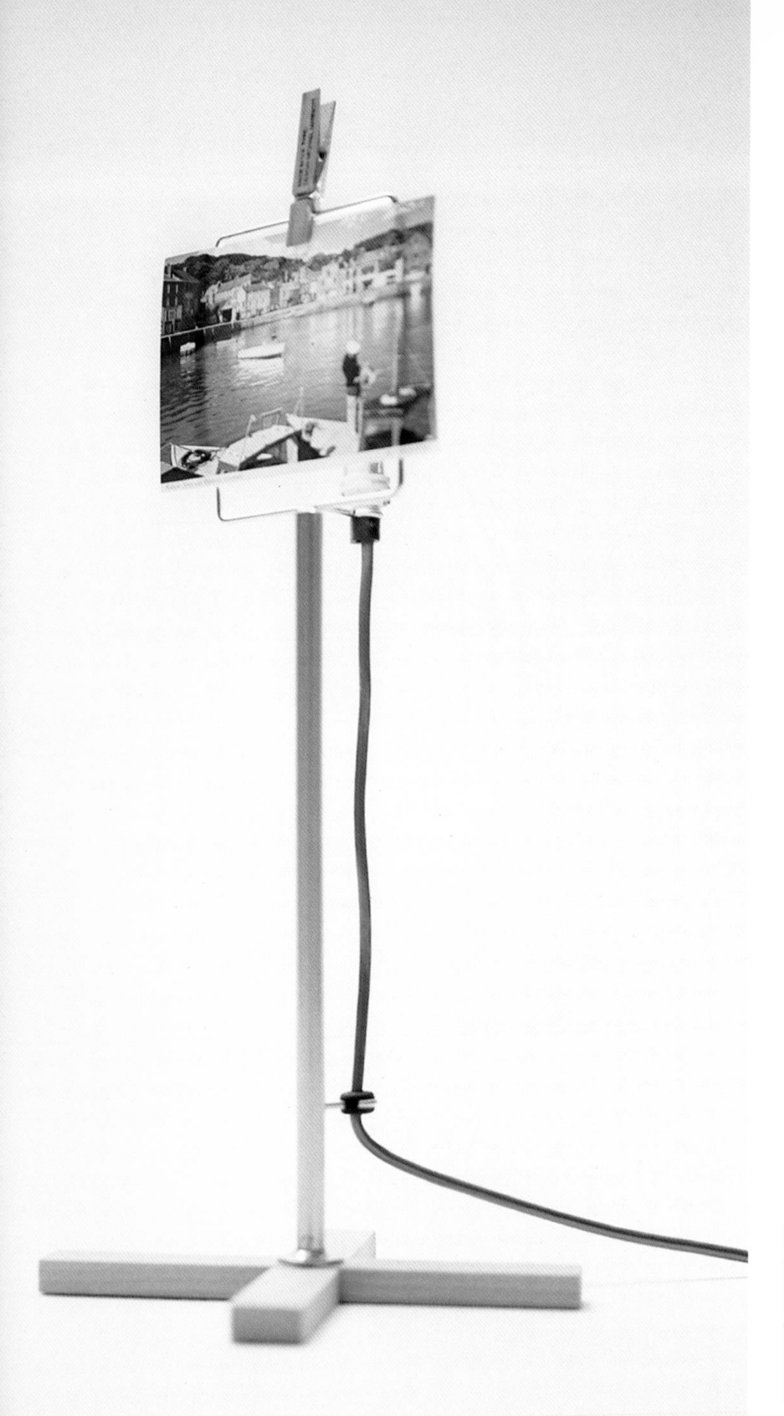

source while shafting the light up to reflect off the ceiling and washing it down to pool on a horizontal surface.

Glowing table lamps can also be used to provide diffused ambient light to a room, although again it is best if they are used at the edge of a room so that some of their light is reflected back. In this case they can, to some extent, counteract the coloring effect produced by using wall washers with colored walls, whereby the colored walls reflect the light and impart it with their own color.

Left Postcard Light *by Michael Marriott (1993), is a steel tube postcard with a wooden clothes peg and wire frame.*

Above Shogun Tavolo, *designed by Mario Botta in 1986 for Artemide. Here is a table lamp with revolving perforated sheet-metal diffusers that refer to the cylindrical forms found in Botta's architecture. It is painted white and black and gives a wonderful play on shadows and light.*

Left Fibre Space, *manufactured by Mathmos, has an aluminum base. The fountain of light emits from the revolving fiber optics.*

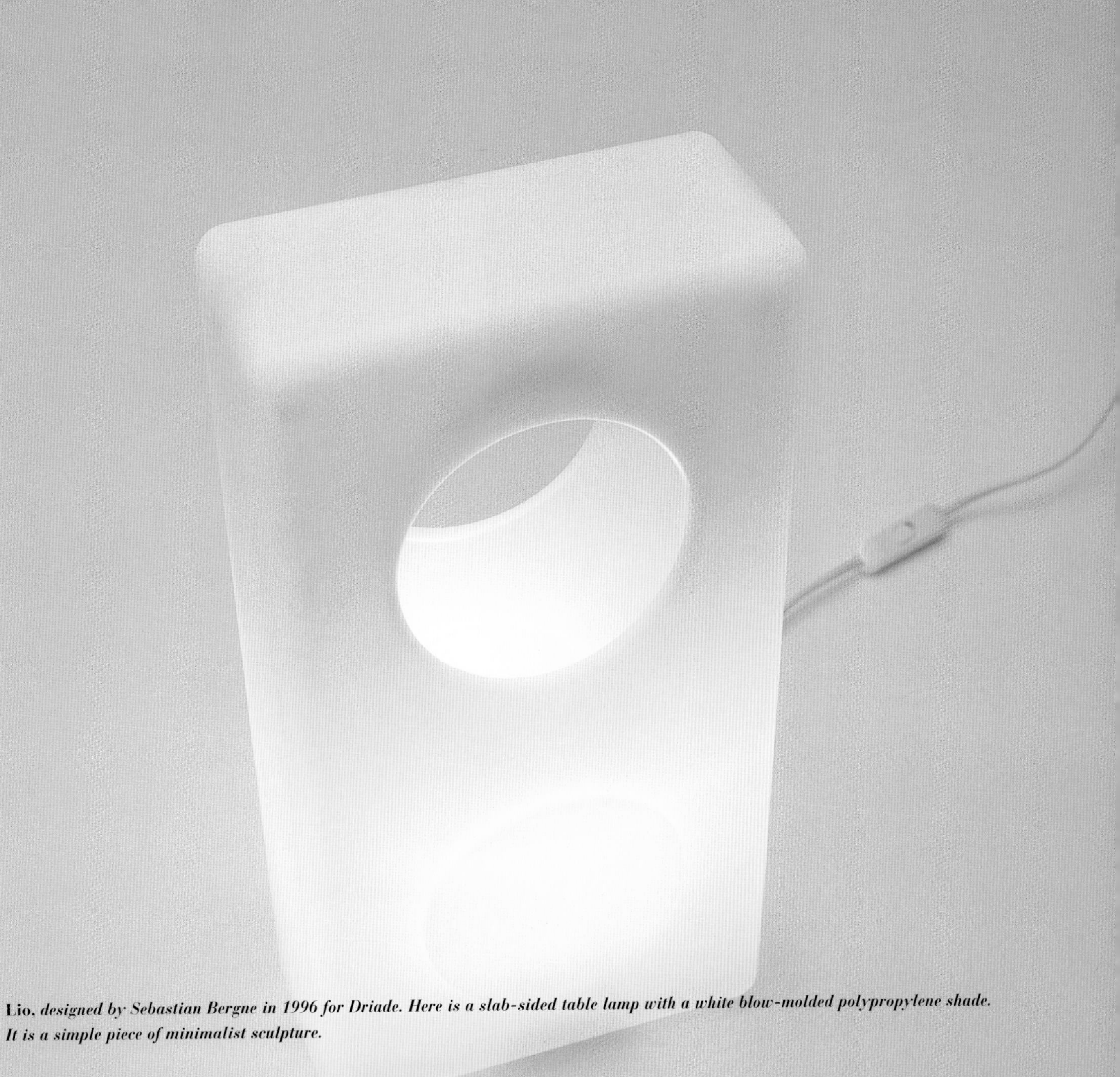

Lio, *designed by Sebastian Bergne in 1996 for Driade. Here is a slab-sided table lamp with a white blow-molded polypropylene shade. It is a simple piece of minimalist sculpture.*

Again, their position in the room and the height at which they are placed will have a profound effect on the quality of light they provide.

Table lamps are versatile but they do need to sit on tables or other horizontal surfaces, which can make them less than versatile to use. It's good to position them near windows so that the light pattern does not change too much with nightfall; always try to position them so that their bulbs are not visible, both from standing and sitting positions. Be sure to avoid having trailing cords and cables—they are not only unsightly but also dangerous if tripped over, causing a mass of ceramic, glass and white hot metal to arc through the room.

Finally, although their primary function is to provide a source of light, some lamps are so inept at this that they become almost purely decorative objects. However, this is clearly a major part of their purpose too. The once reviled but now revered lava lamp is an excellent example of this.

Opposite Solo Table Light, *designed by Douglas Bryden, Richard Smith and Stephen Young in 1998 for the Arkitype Design Partnership. This table lamp comes as a polypropylene kit ready for home assembly. It is a simple and inexpensive way to make a fun light.*

Left Jet, *manufactured by Mathmos in 1999, is made from melamine, spun aluminum and glass. The heat from the lamp causes the colored liquid to blob about in a colorless fluid. Once reviled but now revered, this is probably the most sophisticated contemporary version of the generically labeled lava lamp.*

desk lights

Traditionally, the primary criterion for desk lamps is that they should be useful for reading and writing. However, the proliferation of computers has meant that there is also the additional requirement of providing shadow-free ambient light for the keyboard and to counteract the glare of the screen, which otherwise can be quite tiring. Again, the physical mechanism is to reduce the pupil size, increasing usable focal distance. It is useful to position these lamps next to anywhere you sit to read or next to a telephone where you might need to take messages. They can also be used to provide localized spotlighting, such as pools of light for floral displays or art objects.

These lights need to be practical and easily adjustable—good examples are the seminal *Anglepoise* by George Carwardine (see page 17) and Richard Sapper's equally innovative *Tizio* (also see page 17). The *Tizio* heralded the introduction of the tiny, cool, quartz-halogen capsule lamps that gave this genre of lights the benefit of smaller lamp heads. These don't get in the way, and they allow a lighter, more elegant structure to be designed. The high-tech mechanical articulation inherent in these

Berenice, ***designed by Paolo Rizzatto and Alberto Meda in 1985 for Luceplan. This lightweight insect-like form is one of the most beautiful desk lamps of the genre. Made from die-cast aluminum and stiffened nylon, the pressed glass reflector has an enameled white interior and UV-protection glass.***

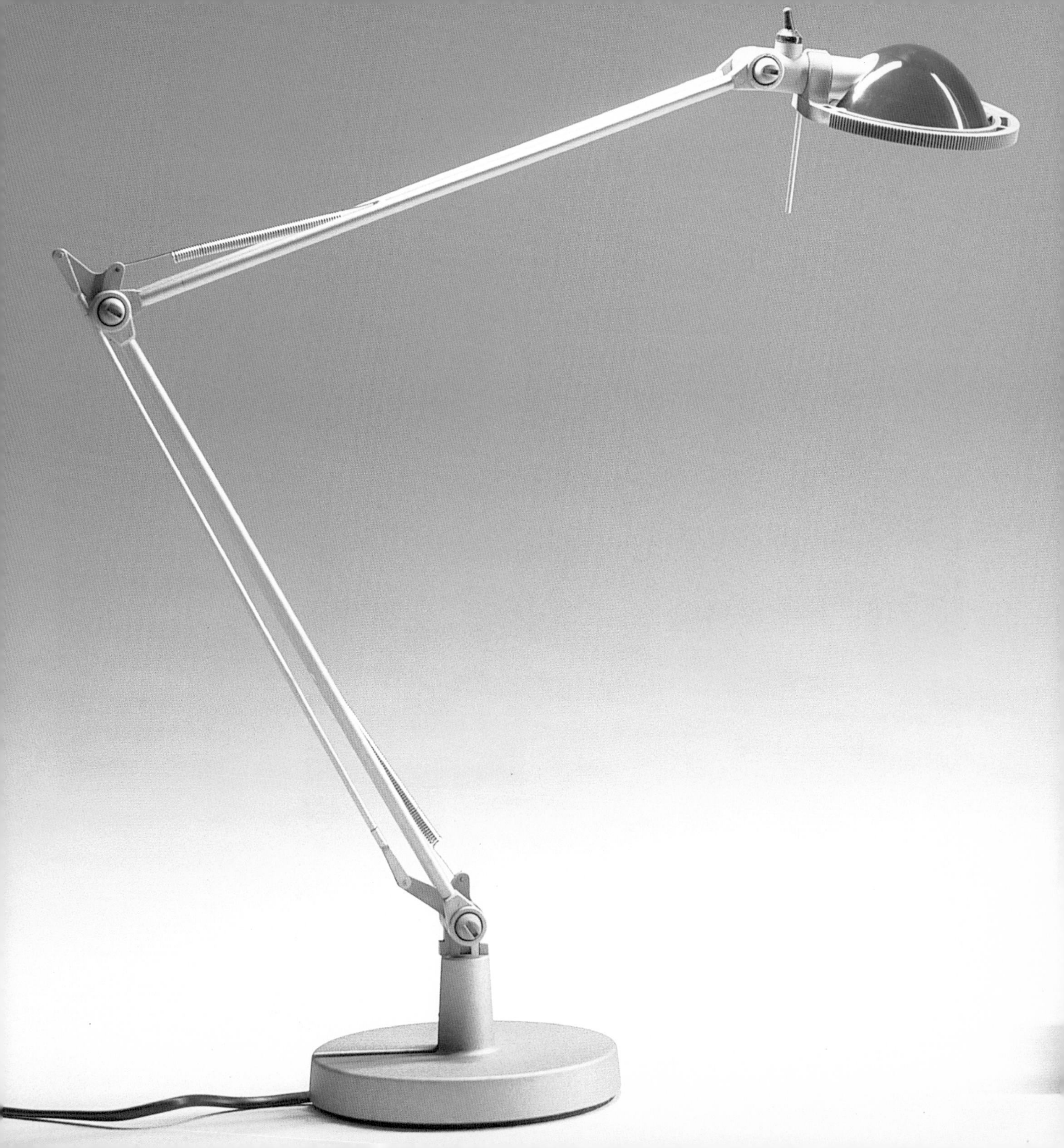

lights often gives them an insectlike lightness similar to that of a grasshopper or dragonfly. Desk lamps can also have the feel of miniature building cranes, fire truck ladders or cherry pickers—I wonder how long will it be before we see someone design one using just such a toy for the mechanism?

Bedside tables are another habitat for desk lamps. Bedrooms must feel inviting and a key to feeling cozy is switchable lighting very close to the bed. Bedside lamps can be aimed at pillows and are indispensable for reading in bed. Whether wall- or table-mounted, lamps with cantilevered arms can be easily adjusted to move the lamp head over the head of the bed, throwing a bright pool of light on the page.

Below left Fortebraccio, *designed by Alberto Meda and Paolo Rizzatto in 1998 for Luceplan. The chic-cum-macho cantilever mechanism used here appears to be a celebration of precision engineering. It is finished in metal, black, yellow or red.*

Below right Ray, *designed by Caputo and Power in 1996 for Fontana Arte. The lamp shade on this table lamp is suspiciously flowerpot-like.*

Opposite AJ, *designed by Arne Jacobsen in 1960 for Louis Poulsen. A classic from a design guru, the shade of this luminaire can be pivoted vertically. The table light has a push-button switch in the base, while the floor light has a foot switch.*

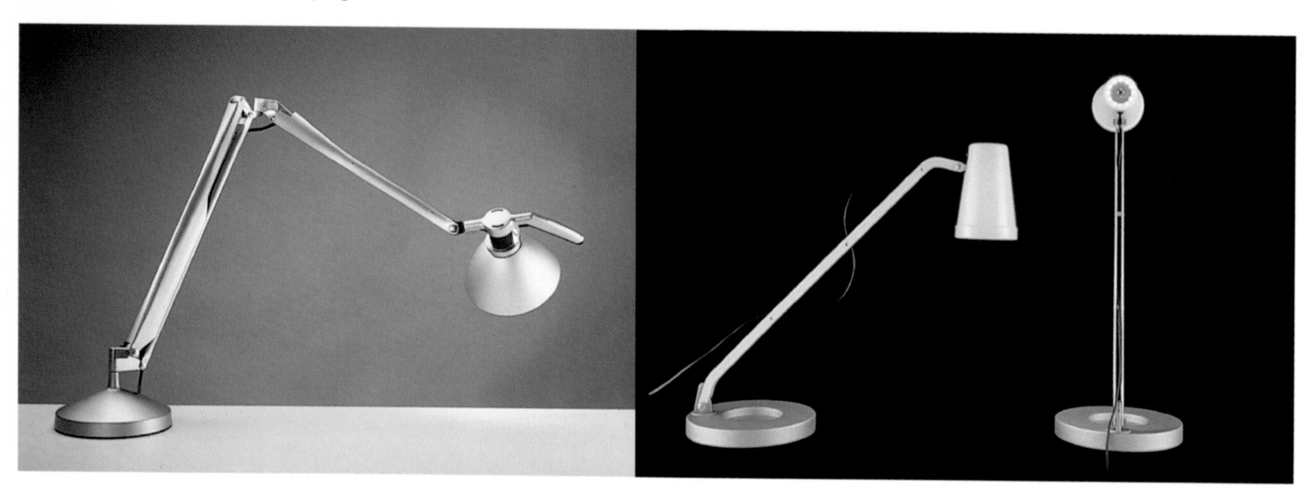

ceiling lights

The chandelier is the queen of all pendant lighting. Used in great halls, the point sources of the candles are enhanced with high-refraction cut crystal, which teases apart the spectrographic components of candlelight into iridescent blues, yellows and reds, producing a rich glittering effect. The candelabra bereft of crystal comes next in the social pecking order, elegantly giving a good light to feast by. Both are akin to floating sculpture radiating from the center of a space.

Pendant lights, so popular in the 1950s and 1960s, have recently and perhaps not surprisingly enjoyed a revival. Depending on their design, they can often combine the virtues of both uplights and downlights in various proportions. Unfortunately, they can emphasize some of the vices too. Sometimes practicality can give way to the preferences of fashion; but life would be pretty boring if it didn't!

Uplighting lifts the appearance of a ceiling, giving the illusion of more space. However, compared to that of uplighters, the shadow a pendant casts will be bigger, sharper and darker. Beware of pendant lights, because, unless used carefully, they can be a disastrous mistake—especially in smallish bedrooms and living rooms. Too close to the ceiling looks harsh, too low means you bang your head! They are undoubtedly at their most effective when used in multiples to light large spaces.

Star, *designed by Tom Dixon in 1997 for Eurolounge. This striking pendant light can also be used as a table or floor lamp. It was allegedly inspired by styrofoam cups stuck to a child's ball and is manufactured from rotationally molded polyethylene, available in a selection of colors.*

Right *Pendant and table lamp from the* Terra Lighting Range, *designed by Sebastian Bergne in 1997 for Proto Design. More slip-cast ceramic forms are shown on pages 58–9.*

Below Paper Pendant, *designed by Nazanin Kamali for Aero. This cylindrical pendant shade in white, with a mild-steel frame in silver or black, has a simple shape with neat joining detail and cord. It also packs flat.*

Overleaf Fucsia, *designed by Achille Castiglioni in 1996 for Flos. This cluster of glass cones is a modern interpretation of the chandelier but is even more serene and beautiful. As shown, it is available in several groupings and configurations.*

Remember that if you hang anything from the wall or ceiling it makes the room seem smaller, so be sure that you have sufficient ceiling height before committing to your purchase. The situations where this type of light work best are high hallways and over dining tables—where they light the food better than the people eating it. This problem is somewhat alleviated when the lights are used in multiple batteries. Again, for this reason, pendant lights do not work well in living rooms.

Small hanging pendant lights with translucent shades can be used above work surfaces, producing beneficial small pools of light for working. However, it is vital that they are positioned so that they don't create shadows or produce glare. The kitchen is also an area where ceiling lights are used quite frequently, often to good effect. However, consider what the lamps will look like with a thin film of grease on them. Don't situate them near your stove, especially if they are going to be a pain to clean—a dusty, greasy pendant is a miserable sight.

Top left Cina, *designed by Rodolfo Dordoni in 1994 for Flos. The molded glass shades are reminiscent of traditional Japanese paper lanterns and use conventional 150W bulbs.* **Top right** Anywhere Light, *produced by Aero. Here, a spun aluminum, bobbin-shaped shade spreads light up and down.* **Bottom left** Daisy (1998), *made from polypropylene and also available as a floor-standing light, was designed by Roy Sant.* **Bottom right** *The shade of this* Filo Pendant, *designed by Peter Christian in 1994 for Aktiva, is made from polycarbonate in white, orange, green and purple* *(see also page 54).*

Above Light Light, *by Clementine Hope (1998). These plastic images of a chandelier have been screen printed on semi-opaque plastic panels with a water-based ink. Available in white or gray; the plastic panels fit together to form a cube-shaped lamp shade that readily attaches to a pendant light-bulb holder.*

Opposite top left UFO, *designed by Nick Crosbie in 1996 for Inflate. This pendant light is inflatable and made from vacuum-formed PVC.*

Top right UFL, *designed by Peter Wylly in 1998 for Babylon Design. Here is a pendant light made from aluminum with polypropylene: the multicolored filters give off a warm-hued glow.*

Bottom left Alzaia, *designed by Vico Magistretti in 1996 for Fontana Arte. This large (it has a diameter of 20 inches) but elegant pendant shade with translucent glass lamp holder houses a hefty 150W conventional incandescent bulb.*

Bottom right Zuuk *by Ingo Maurer (1994) has a stainless steel, colorless, heat-resistant, satin-frosted glass halogen bulb. It comes complete with a surface-mounted ceiling fixture and anodized base-plate cover.*

The general effect of downlighting is that of visually dropping a ceiling—because the ceiling is in shadow it therefore appears to come closer to the viewer. Whether track-mounted or recessed, ceiling lights are especially effective in halls and corridors where they illuminate the floor and pathway. They also can be angled as wall-washer floods or spotlights for pictures and architectural features. If used as wall washers they provide effective curtains of light, too. Positioned carefully over seating, they can also provide a bright light by which to read books and magazines.

Left Waves, *designed by Johannes Norlander in 1998 for Box. The opaque shade with its large diameter and stubby "fingers" gives the curious effect of an upside-down splash of milk. Note the adjustable coat-hanger-hook detail above the shade.*

Right and opposite Saturn Lamp, *produced in 1999 by Jam. The housing of this lamp is formed using closed-cell polymer foam and an Ecotone bulb, which infuses the translucent foam with a brightly colored soft organic light.*

CITY
SPICE
OPEN

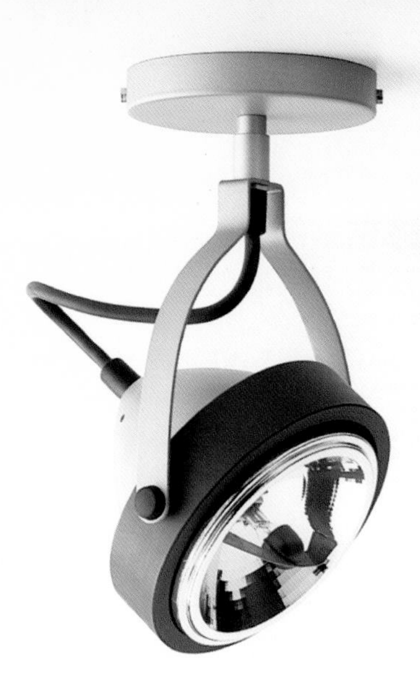

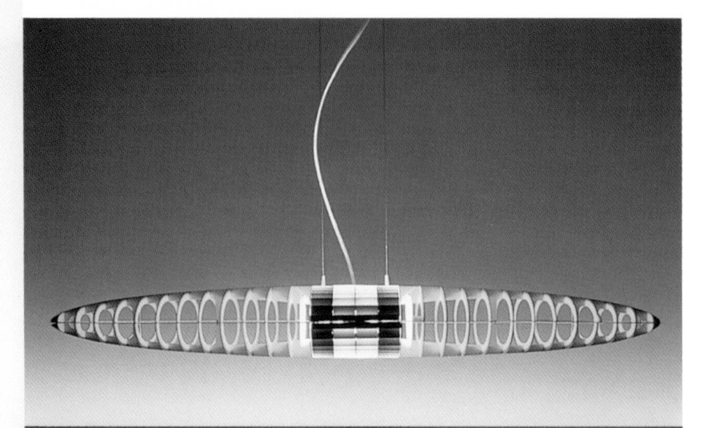

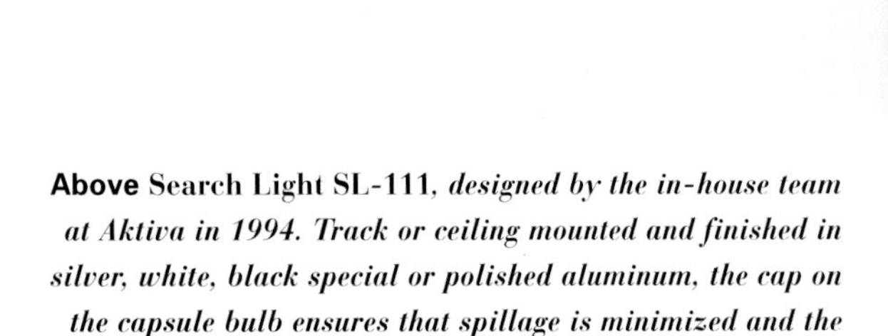

Above Search Light SL-111, *designed by the in-house team at Aktiva in 1994. Track or ceiling mounted and finished in silver, white, black special or polished aluminum, the cap on the capsule bulb ensures that spillage is minimized and the light is focused in the direction intended.*

Top right Titania, *designed by Alberto Meda and Paolo Rizzatto in 1995 for Luceplan, featuring a lamellar shell in natural aluminum. Five pairs of interchangeable polycarbonate filters, silk screened in green, red, blue, yellow and violet, determine the coloring of the shell while still continuing to emit a white light. Redolent of a skinless air-frame, the skeletal structure has an eerie, UFO-like presence.*

Right Romeo Moon S2, *designed by Philippe Starck in 1998 for Flos. With molded-glass or fabric shades, this pendant light is part of a system that offers coordinated lights for table, wall and floor use. A semi-industrial looking object, it has been house-trained for domestic use.*

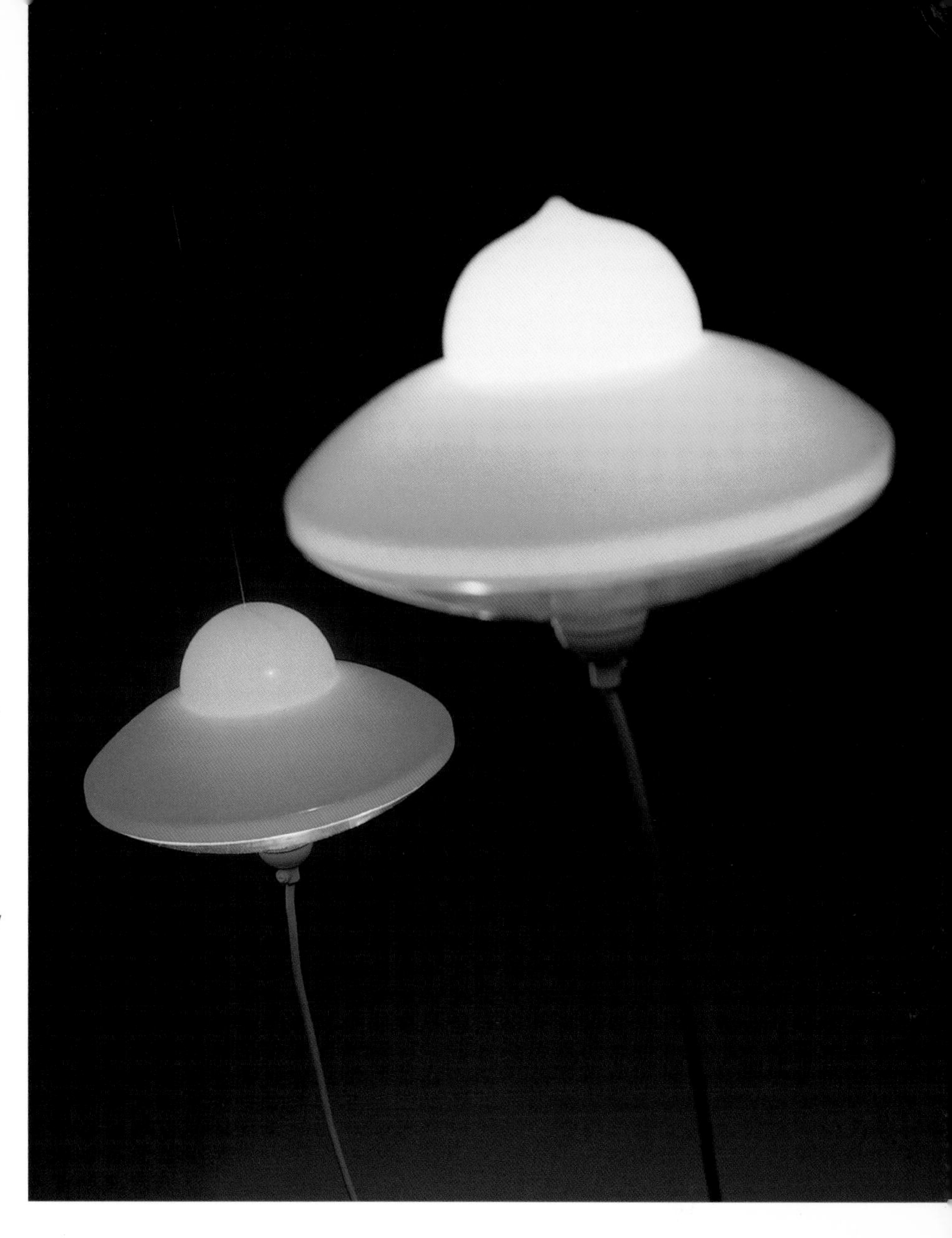

Above Clio, Erato, Urania. Musa sospensione, *designed by Rodolfo Dordoni in 1994 for Artemide. Suspension lamps with and two- or three-colored glass diffusers are available in many different versions, utilizing the components in different configurations.*

Right Don't run we are your friends, *designed by Roberto Feo for El Ultimo Grito. In these adjustable pendant lights made of acrylic plastic, the thin connecting wires coming from the tip of each shade are passed over ceiling-mounted eyelets, giving a seesaw effect to the deliberately UFO-looking lights.*

Glo-Ball Pendant, *designed by Jasper Morrison in 1998 for Flos. Shaped like a tangerine, this ceiling light is designed in the spirit of minimalism.*

Below left Ya Ya Ho, *designed in 1984 by Ingo Maurer and team for his own company, is a low-voltage lighting system fed by a transformer to reduce the house current. It seems to have been inspired by the overhead power cables used in many European tram systems.*

Top right Mikado Track, *designed by F. A. Porsche for Artemide. This ceiling- and wall-mounted track-light system has a series of little conical lights with a high-tech appearance. Such track-lighting systems are particularly useful for highlighting pictures.*

Below right Medea, Erilo Chronocolour series, *manufactured by Artemide. These semi-recessed 12V dichroic lamps have transformers housed in the ceiling void. The small colored diffusers are there to add interest more than to reduce light spillage and glare.*

Opposite left PH Snowball, *designed by Poul Henningsen in 1958 for Louis Poulsen. This a height-adjustable pendant with a series of curved metal screens. The inside is matte, outside has a high-gloss white-finish screen to give an evenly focused, diffused light.*

Opposite right Satellite Pendant, *designed by Vilhelm Wohlert in 1959 for Louis Poulsen. This onion-shaped opal-glass dome is open on the underside to produce an overall glow.*

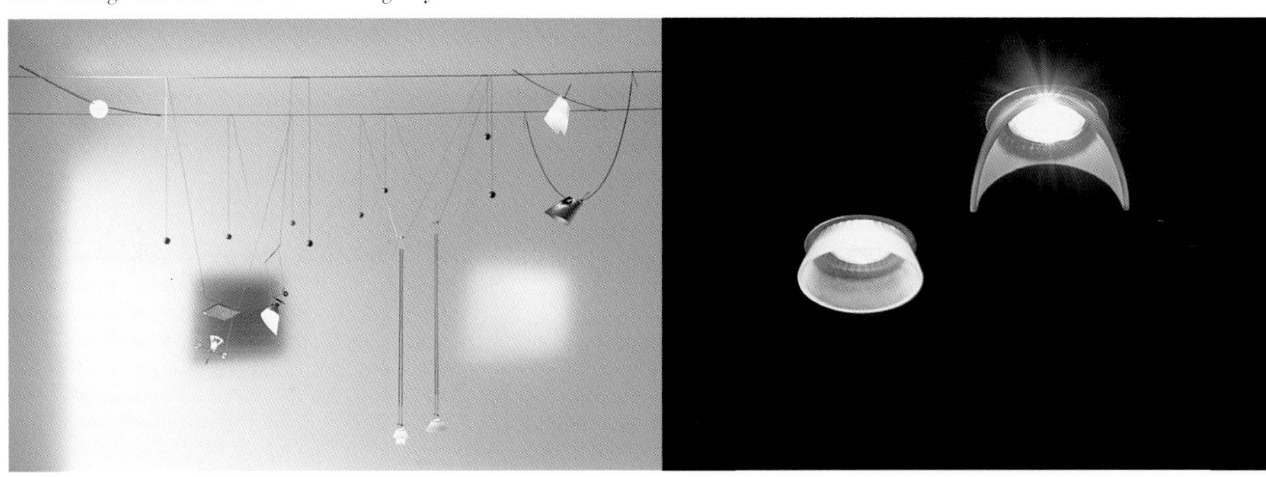

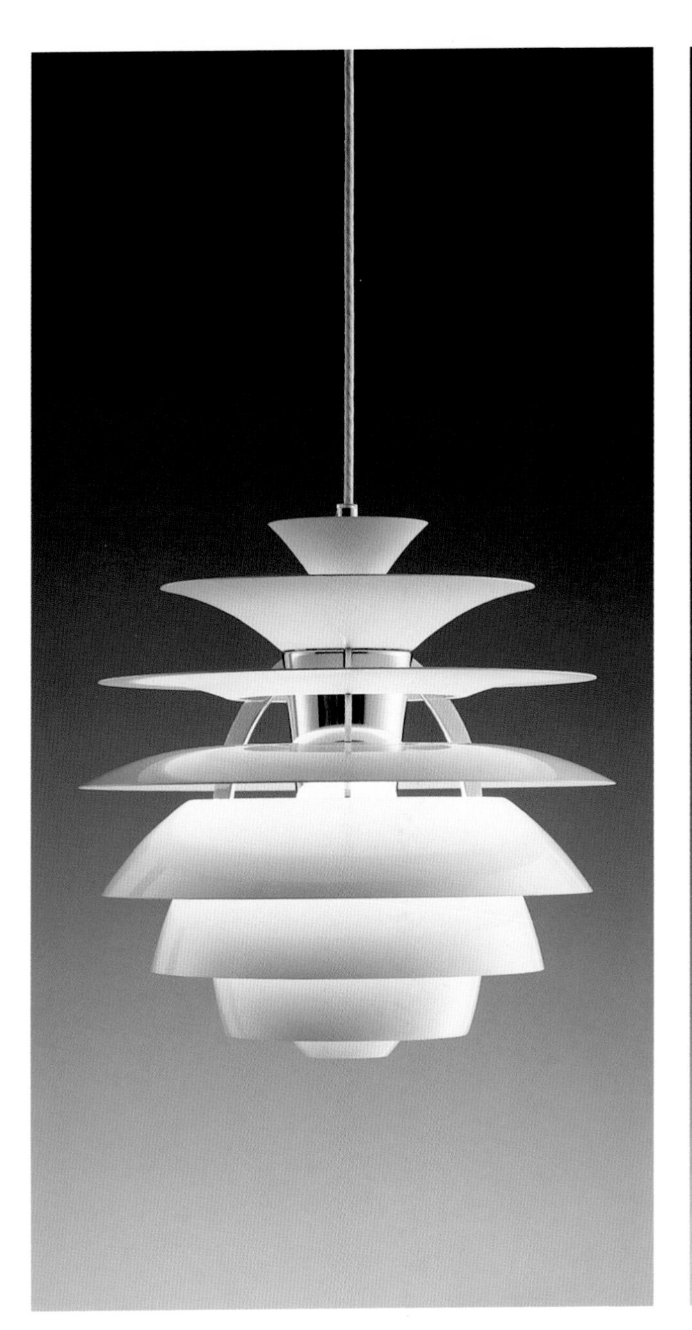

The general effect of downlighting is that of visually dropping a ceiling—because the ceiling is in shadow, it appears to come closer to the viewer.

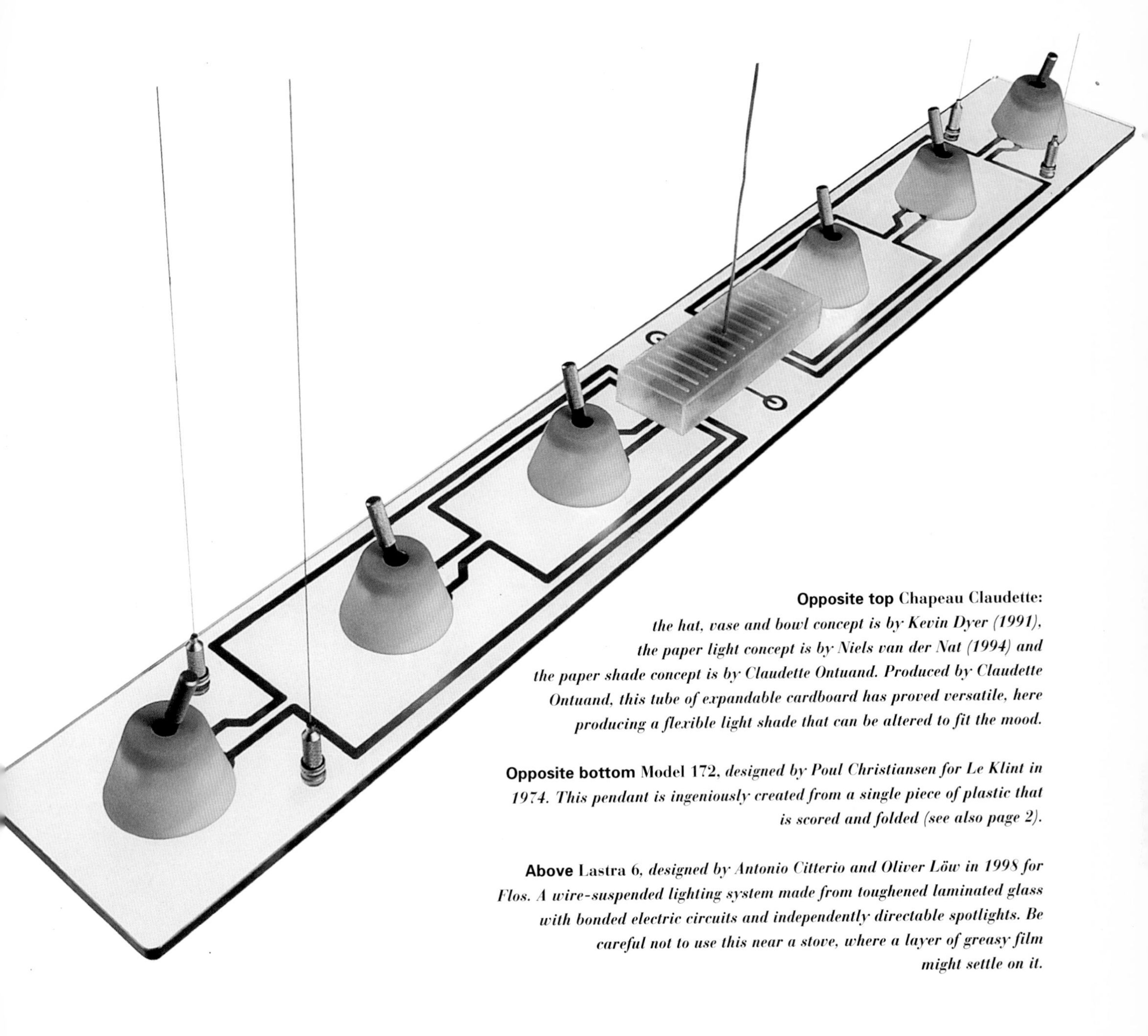

Opposite top Chapeau Claudette: *the hat, vase and bowl concept is by Kevin Dyer (1991), the paper light concept is by Niels van der Nat (1994) and the paper shade concept is by Claudette Ontuand. Produced by Claudette Ontuand, this tube of expandable cardboard has proved versatile, here producing a flexible light shade that can be altered to fit the mood.*

Opposite bottom Model 172, *designed by Poul Christiansen for Le Klint in 1974. This pendant is ingeniously created from a single piece of plastic that is scored and folded (see also page 2).*

Above Lastra 6, *designed by Antonio Citterio and Oliver Löw in 1998 for Flos. A wire-suspended lighting system made from toughened laminated glass with bonded electric circuits and independently directable spotlights. Be careful not to use this near a stove, where a layer of greasy film might settle on it.*

wall lights

Historically, wall lights stem from the days when flickering, flaming torches were de rigueur, attached to the wall with wrought iron fittings. But with the advent of candles and more delicate oil lamps they lost favor a little—at least until the invention of piped coal gas in the nineteenth century, when wall lighting became particularly popular once again. Recently, the trend has been not only to use them to provide direct light but also to act as floodlights, reflecting light off the ceiling into rooms. This has been particularly successful with the lower ceilings in contemporary environments.

Ariette, *designed by Tobia Scarpa in 1973 for Flos. This kite-shaped wall or ceiling lamp gives diffused light, housed in fire-resistant synthetic fabric stretched over a sprung framework. It is available in three sizes and uses a gang of four 40W incandescent golf-ball lights.*

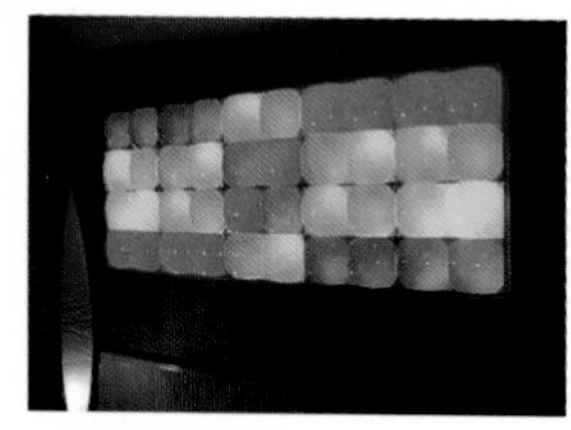

Left and right Chromawall, *designed by Jeremy Lord in 1995 for the Colour Light Co. Chromawall is a modular, wall-mounted changing light display system. Each module radiates light that changes color slowly and smoothly. There are no moving parts and the colored lamps are dimmed up and down by electronic controls. Each module contains four color change cells, which each include four specially colored filament lamps.*

Below left Wall A Wall A, *designed by Philippe Starck in 1993 for Flos. This is a diffused-light wall lamp with a colored wall bracket and a milky-white plastic diffuser with colored filters. Its ingenious switch gently glows when the light is off.*

Below right Train Train, *designed by Marc Sadler in 1996 for Flos. The ability to evenly light the face makes this versatile modular light system perfect for a mirror surround. It also incorporates an outlet for electric razors and toothbrushes.*

Opposite, clockwise from top left Drop 2, *designed by Marc Sadler in 1993 for Flos. The lamp cover is molded in a translucent, flexible compound. The base is a rigid translucent plastic that gives a colored glowing halo.* Icon "Holes," *designed by Peter Christian in 1996 for Aktiva. It has a screen-printed image and an anodized aluminum frame that comes with several different patterns and in various panel sizes.* Lola Wall, *designed by Alberto Meda and Paolo Rizzatto in 1987 for Luceplan. With a swiveling carbon-fiber head, micro-perforated metal reflector and protection glass, this organic-looking piece will take a 250W halogen bulb and uses advanced technology. The general direction of the light is controlled by moving the small protruding rod.* Search Light, *designed by the in-house team at Aktiva in 1994. A wall-mounted uplighter available in two versions and different finishes—silver, white, black or polished aluminum.*

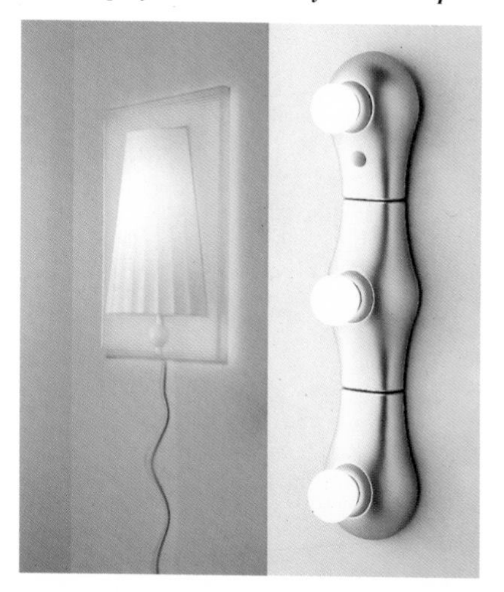

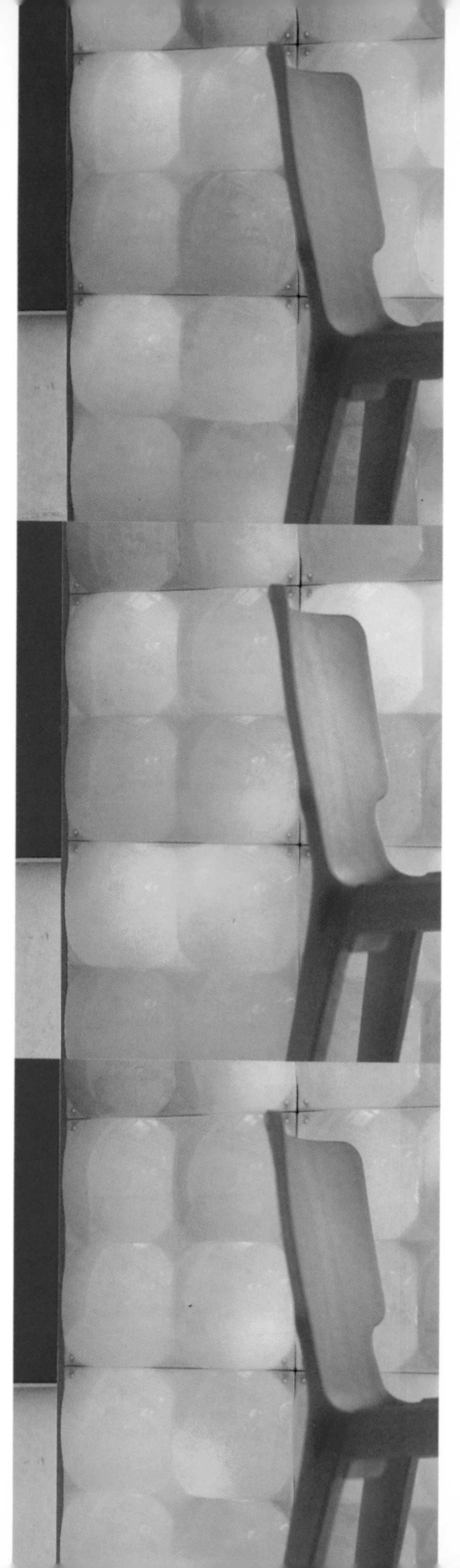

There are generally three types of wall lighting—floods, which cast light back onto the wall or ceiling; glowing, which emit a diffused light directly into a space (or a combination of the two); and spotlights, which project bright, tight pools of light with sharp shadows. An entry hall is a place for display and the natural habitat of wall lights, since they are off the floor and out of the way. In addition to providing a friendly introduction to the home, wall-mounted lights provide a dramatic focal point to a smallish space. The lighting level should generally be fairly subdued in preparation for entering the actual living spaces, but make sure that there is adequate light on the floor for a feeling of welcome and security.

Top Circus Grande, *designed by Defne Koz in 1994 for Foscarini. The large, convex bun-shaped plastic molding stands slightly away from the surface to give good ambient light. This light is available in two sizes and can be used equally well on the wall or the ceiling.*

Center Screen, *designed by Alberto Meda and Paolo Rizzatto in 1989 for Luceplan. This is an injection-molded acrylic screen with a white silk-screened prismatic surface. It is available in two sizes and works particularly well near mirrors, where the large area of even light is especially effective.*

Bottom Acheo Wall, *designed by Gianfranco Frattini in 1989 for Artemide. This industrial-looking painted-metal and Pyrex glass floodlight comes with various wall and ceiling fittings.*

Zero One, *designed by Ingo Maurer and team in 1990 for his own company. A white Corian body with frosted mirror supports a protective glass cover that surrounds a halogen bulb. The light from this brick-sized light can be easily adjusted by moving the reflecting, diffusing mirror along the wire support.*

Rubber Light, *designed by Mark Bond in 1998 for Bond Projects. The heatproof elastomeric sheath/shade pulls firmly over the light bulb, which nonchalantly hangs from a wooden peg that is screwed into a wall.*

outdoor lights

Fire, and particularly fireworks, provide the most dramatic and spiritual forms of outdoor light. There are many tools for outdoor lighting: floodlights, pendants hung from trees, miniature Christmas tree lights and firelight—but the greatest of these is always flame. It's hard to beat the romance and carefree atmosphere of long, warm summer evenings. This reverie can be prolonged with some sensitive and imaginative exterior lighting. The important thing, however, is to avoid the glaring, floodlit football field approach.

There are two basic types of natural-looking outdoor lighting: plant light and moonlight. Lights hidden in plants make them glow with a mysterious luminescence. Unfortunately, this is no help for general navigation in the dark. For this you need some downlights—as high as possible—to bathe a lawn or terrace. Alternatively, there are bulkhead lights and the waist- or foot-level guiding lights used along public stairways and walkways, but more elegant and miniature ones are available for domestic use.

A more romantic approach is to line paths with long-lasting votive candles in windproof glass holders. These can be simply made by placing a votive candle in a glass tumbler inside a white paper bag—the effect is quite magical: almost like an aircraft runway for the fairies.

Left *In sparklers, steel filings are mixed with an oxidizing agent such as potassium perchlorate and combined with gum arabic. When lit, the mixture spectacularly oxidizes in the way a lamp filament would do if it were not protected from oxygen by a vacuum.*

Right Jack Light, *designed by Tom Dixon in 1996 for Eurolounge. Available in a range of colors, these rotationally molded (like traffic cones) light-emitting nodules can be used in a variety of ways, from stools to table bases. They can also be stacked and are suitable for occasional outside use.*

The urban environment has become a miasma of light—although brash and exciting, this also means that the stars are completely obscured. Above, the reality of modern-day Tokyo eclipses the portrayal of the urban future as seen in Ridley Scott's movie Blade Runner.

In domestic settings, outside lighting tends to be used infrequently, perhaps for summer entertaining and on special occasions. This gives a unique opportunity to be quite radical and theatrical, using colored floodlighting and spots. Alternatively, there is the tasteful, romantic approach with soft, glowing, subdued areas to balance out the brash. Whatever you try, it's a wonderful fact that with a little imagination and daring the most dreary yard or garden by day can become the fantasy of your choice by night. Keep the lights low and use lots of color, with accents of flame and candle.

Architectural details can be dramatically enhanced and revealed with the use of both spotlights and floodlights. Water will always respond enchantingly to some well-focused spotlights. Don't kill the mystery of darkness, though. Rather, be theatrical, using pools and little twinkly areas of light. Christmas tree lights work wonderfully on summer evenings, strung through the odd bush or just lighting the edge of the path. Also, think candles, candles, candles. Enough is not a word that applies to candles—like daffodils, they work best in clusters.

Opposite, bottom left Pantarei 300 Halflight, *manufactured by Artemide. This exterior bulkhead light can be mounted so that the lamp shines down a path or sideways into a portico. It is both discreet and resilient.*

Opposite, bottom right Pod Lens, *designed by Ross Lovegrove in 1998 for Luceplan, with injection-molded polycarbonate, integral prismatic lens and overmolded sides. Primarily intended as an outdoor light, the material is resistant to the elements. The light comes in various colors and fixtures with different lengths of stem and base configurations.*

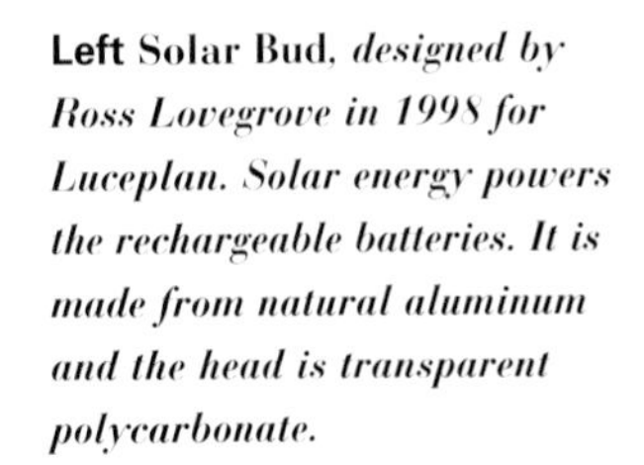

Left Solar Bud, *designed by Ross Lovegrove in 1998 for Luceplan. Solar energy powers the rechargeable batteries. It is made from natural aluminum and the head is transparent polycarbonate.*

Right *Post lamps, or column lights, are good for path illumination, especially if there are no fixing walls available. Usually very sturdy, they provide good perimeter security too.*

The great thing about outdoor lighting is that it can be temporary, so the opportunities for creativity are virtually boundless. Flaming torches, for example, can be used to good effect. Too little is often better than too much, if all you are trying to do is guide and flatter. Trying to provide even, ambient lighting, or using high-level bright floodlights outdoors will all too easily make your garden look like the yard of a minimum-security prison.

room effects

The quality of light is directly related to quality of life. By day or by night the effect of light has a profound—if subconscious—effect on the perceptions of space in any environment.

Good lighting can make the prosaic palatial. Any skilled lighting designer can, I guarantee you, turn the most humble of environments into a dramatic pièce de résistance. (Domestic practicality may be another issue, however.) Experience and skill can make a graveyard capable of seeming like a fairground and vice versa.

Light can relax, enhance and stimulate, but as with all good things too much is as bad as too little. The quality of light can be directly related to quality of life. One has only to think of summer as opposed to winter, where the principal difference between the seasons is, in fact, light driven. Light also has a profound effect on the perceptions of space in any environment.

There are a few simple rules for lighting, and many tricks—the only real sin is to expose a naked bulb (although nakedness does have a place!). If you look and try to understand how its magic works, you can produce truly amazing results for surprisingly little effort.

The surroundings will inevitably color the light reflected off them. Equally, coloring the light will color the surroundings. An all-white room with white light reflects a large amount of evenly diffused light, with the consequence of practically eradicating all shadows and making subtle tonal plays throughout the room. Difficult to keep clean, maybe, but not easy to lose things under a sofa or in a dark corner. The most noticeable impact is the maximization of the illusion of space.

An all-white room also reduces visual clutter and provides a neutral environment where people or art introduce the color. Small, colored low-output lights and bright-hued objects stand out dramatically, chiming with each other where they might have been lost in a busier space.

Another advantage of the white environment is that it's easy to match colors; an interesting and dramatic way to add contrast is to have a strong primary-colored or black bedroom or bathroom (but remember not to use reflected light in these circumstances).

Diffused light shone through a textured surface increases interest and reduces the blandness of the light. Heavily textured walls obliquely lit from the side or from down low make coarse brickwork look interesting.

Putting a standard lamp into a corner works well in white or pale rooms since the light bounces off the walls, enhancing the illusion of space. In bedrooms it is essential to have some carefully-positioned light sources that can illuminate anything you might be reading.

Placed centrally in a room, a glowing object can provide useful localized ambient light. It can be especially effective when used down low, in corners, next to walls or on tables. It's important to consider what a space will look like both in daytime and at night and to plan the lighting accordingly—always consider what effect a lighting fixture will have on its environment when it is turned off as well as when it is turned on. Sculptural forms that reflect light during the day and emit their own light at night enhance their surroundings in either mode—lights can look good while they are supplementing daylight.

The use of light is as much about the use of darkness and shadows, which give form and drama, as it is about the stuff of light itself.

Color is a very important issue too; keeping a controlled color scheme will give any environment a strong identity. Color effects can be supplemented by a subtly tinted light, which gives added interest to the colored objects and walls. Avoid bright primary colors unless you are looking to decorate with a toy store or nursery effect.

Semi-opaque lamp shades hide the glare of the light source while channeling the light up to the ceiling and down onto the horizontal surfaces, producing pleasing, glowing pools of light that can add warmth to a room full of otherwise subdued light. This effect is why this type of lamp shade is used so often in reception room environments; the trick is to make it look contemporary—not like the abode of an aged aunt.

Small lights make excellent and thoughtful gifts for many occasions. Scented candles have traditionally made good gifts to bring when visiting someone's home. They are a pleasure to receive—you don't have to rush off to arrange them like a bunch of flowers—and easy to pass on to others in a pinch (an ancient and revered tradition).

Small, quirky lights add charm and give pleasure. Little lights dotted about like tiny dancing fairies can fill out and add interest to low-level ambient light. Candles and tea lights are especially useful for this—the primary purpose of these lights is to add a little zest rather than illumination with a practical use. When the predominant ambient light is turned down low they can spring out, becoming much more apparent.

contemporary

designers

achille castiglioni

Achille Castiglioni could be said to be the godfather of contemporary lighting. He was born in 1918 in Milan, Italy, the youngest of three brothers, all of whom became professional architects and designers. He started off designing in partnership with his elder brother Pier Giacomo and together they had a strong influence on the Flos lighting company from its foundation in 1962. Flos was one of the first companies to use low-voltage automotive-type lamps for domestic lighting, with the seminal *Toio* telescopic uplighting floor lamp (see page 45).

Achille Castiglioni's work communicates ideas in a charming way. Timelessness, innovation, lateral thinking, wit, ingenuity and unpredictability are ingredients in all his work. This said, unlike the work of many designers, there is no discernible personal style running through Achille Castiglioni's designs—instead, each piece has its own strong individual character.

Brera S, *designed for Flos in 1992. This simple and elegant blown glass and plastic construction uses an incandescent bulb. This lamp head can be used in a variety of permutations: pendant, wall, ceiling, floor and table.*

Left Diabolo, *designed for Flos in 1998 and made of spun aluminum. This pendant light hides almost 5 feet of cord in the upper cone and can be pulled down to extend as needed.*

Below middle Noce, *designed in 1972 for Flos. This gigantic, oversized bulkhead light has an enameled aluminum base and pressed clear-glass diffuser.*

Below right Luminator, *designed by Achille and Pier Giacomo Castiglioni in 1954 for Flos. This floor lamp uplighter gives indirect light and was one of the first fixtures designed around the integrated reflector mushroom spot lamp. The design remains fresh, even today.*

Left Arco, *designed by Achille and Pier Giacomo Castiglioni in 1962 for Flos. Here is a portable pendant floor lamp that gives direct light where it's needed. The white marble base has a hole in it directly above the center of gravity so that two people can push a broom handle through to lift it. The lamp has an adjustable telescopic stainless-steel arched stem and polished aluminum adjustable reflector.*

ingo maurer

Ingo Maurer was born in 1932 in Germany. For many years he worked as a designer in the U.S. before returning to Europe in the early 1980s to specialize in lighting design. Ingo Maurer initially explored the opportunities offered by miniature low-voltage dichroic-quartz halogen lamps with his groundbreaking *Ya Ya Ho* lighting system (1984), which had two parallel wires carrying either terminal of a 12-volt circuit (see page 80). Since then he has produced more and more seemingly outrageous but often quite significant design pieces. He has always continued to explore the boundaries of available technology and to incorporate it into his work in increasingly imaginative and ingenious ways.

Many lighting designers have what might be described as somewhat eccentric personalities compared to those concerned with more tangible ideas, and Maurer is best known for his lighthearted and whimsical approach. Often his work is closer to an art installation than design—and even then it's not so much sculpture as light-emitting physical poetry. He does have a clever eye for the practical side too, making designs viable but without neglecting his wonderful childlike imagination.

Left Wo bist du, Edison?, *designed in 1997. When is a light not a light? When it's a hologram. A circle of holographic film catches the light emitted by a dichroic reflector housed in a holder modeled on the Edison bulb profile. The effect is to create a mysterious mirage of a glowing light bulb.*

Right Zettel'z, *designed in 1997 with stainless-steel, heat-resistant satin-frosted glass and Japanese paper. This self-customizable light is made from a kit of parts that invites the user to participate in creating its final incarnation.*

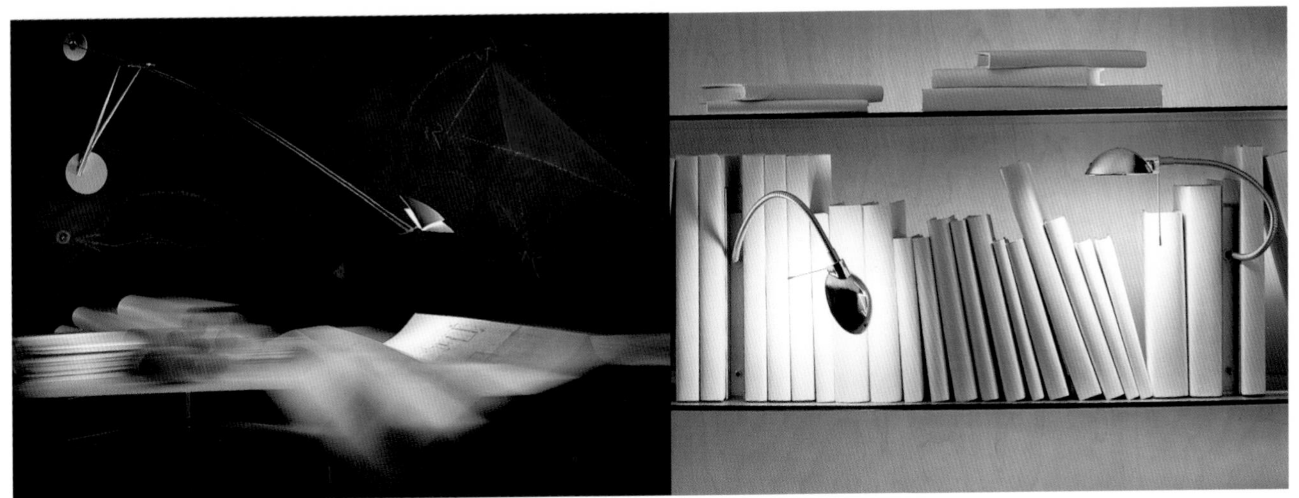

Clockwise, from top left Tijuca Wall, *designed in 1989. Reminiscent of a miniature Sydney Opera House on the end of an articulated fishing rod, these wall-mounted task lights have a good bedside manner.* Oskar, *designed in 1998. Made of anodized aluminum, these electric bookworms neatly nestle among the great authors on your bookshelf. In* Light Structure, *designed with Peter Hamburger between 1970 and 1983, the plastic canopy pays homage to the design guru Buckminster Fuller.* Savoie, *designed with Donato Savoie in 1979. The canopy and socket are made in white porcelain, housing a semi-frosted bulb.* Pierre Ou Paul, *designed in 1996, is a massive, height-adjustable aluminum shade with hand-gilded interior.* Bellissima Bruta, *designed in 1998. This extraordinary piece was unveiled as a unique prototype in 1999 at the Milan Design Fair, a showcase of the latest electronic lighting control technology.*

philippe starck

Philippe Starck was born in Paris in 1949 and the best way to describe this ultra-prolific designer is "mad-genius-showman." Predictably unpredictable, he works in all fields of design including lighting, interiors, furniture, products and graphics—he has even designed motorcycles and scooters. Originally trained as an architect, he is now probably best known for the charismatic forms of his furniture and domestic artifacts, especially his lights.

Without constraint, he seems to have single-handedly created a recognizable, flamboyant, international design style, which combines strong conceptual ideas, functional ingenuity, wit and provocative form with an astonishing attention to detail. Not only does he look for innovative design solutions, he combines them with an exploration of materials and process technology. The boldness of his often irreverent approach together with the sheer volume of his work has made him perhaps the best-known and most noticed designer alive today.

The originality and lack of compromise evident in his work has sometimes meant that it did not gain immediate public acceptance. However, his influence on the design industry is difficult to refute. Perhaps he could be better described as a sort of court jester, given his ability to imaginatively address real design issues in a charmingly lighthearted but gutsy manner.

Below Miss Sissi, *designed by Philippe Starck in 1991 for Flos. This miniature table lamp is made in brightly colored, beautifully molded plastic. It is both inexpensive and useful, giving direct light up and down and some colored diffused light to the side.*

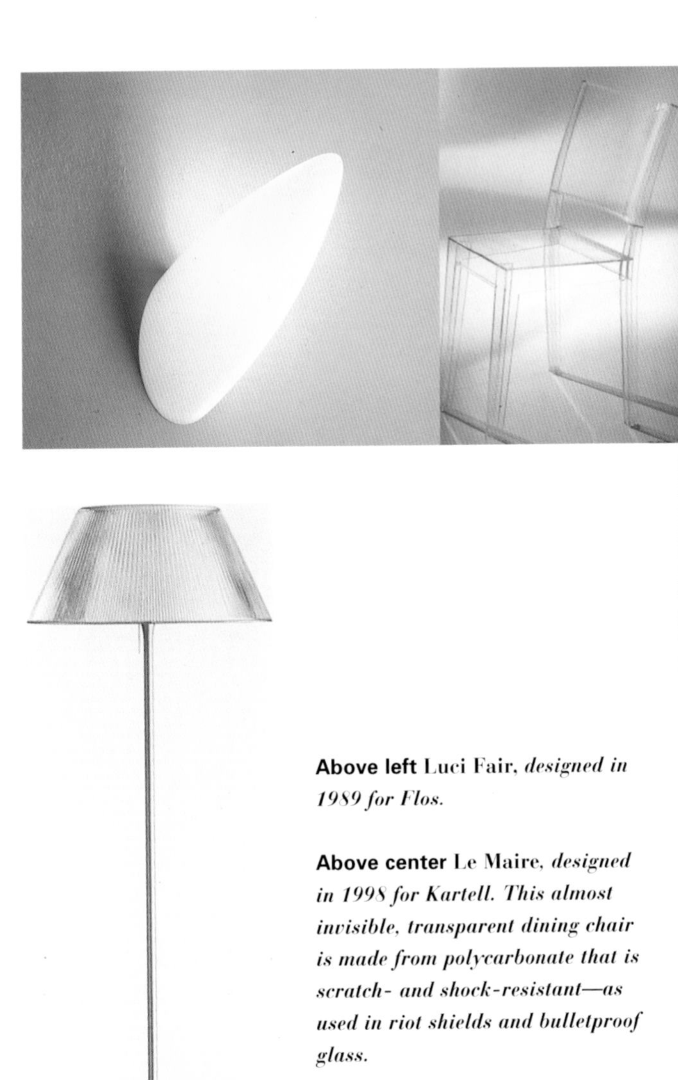

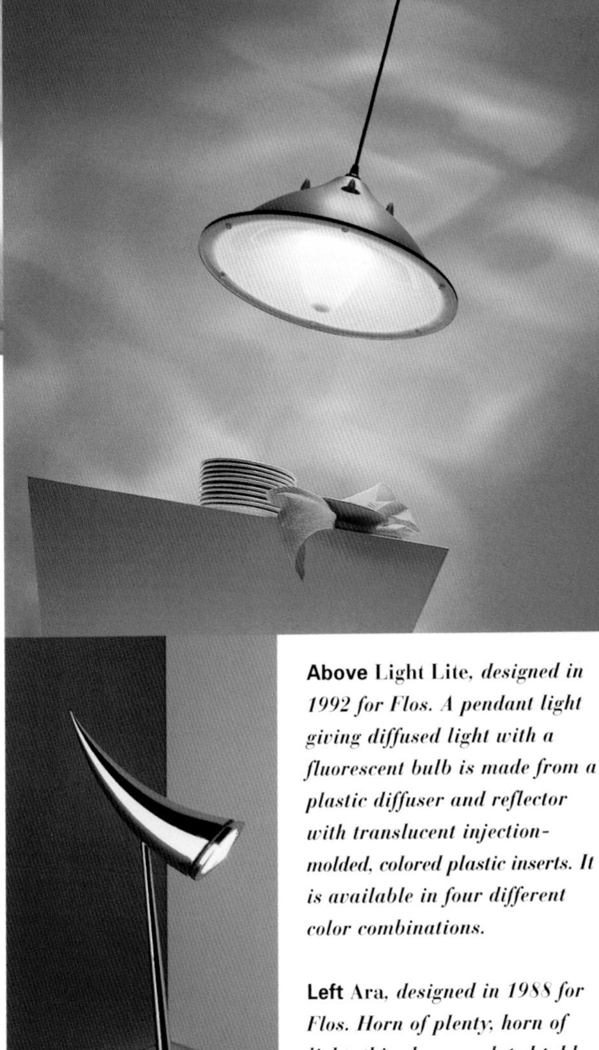

Above left Luci Fair, *designed in 1989 for Flos.*

Above center Le Maire, *designed in 1998 for Kartell. This almost invisible, transparent dining chair is made from polycarbonate that is scratch- and shock-resistant—as used in riot shields and bulletproof glass.*

Left Romeo Moon, *floor version, designed in 1998 for Flos (see page 43).*

Above Light Lite, *designed in 1992 for Flos. A pendant light giving diffused light with a fluorescent bulb is made from a plastic diffuser and reflector with translucent injection-molded, colored plastic inserts. It is available in four different color combinations.*

Left Ara, *designed in 1988 for Flos. Horn of plenty; horn of light, this chrome-plated table lamp has a lens that gives direct light and an integrated tilt switch in the structure that responds to movement in the lamp head.*

tom dixon

Born in Tunisia in 1959, Tom Dixon started adult life as bass guitarist in the soul band Funkapolitan, but after breaking his arm on a motorbike he was unable to swing his arm with quite the same verve. Deciding to become a design star instead, he learned how to weld and in 1985 inaugurated Creative Salvage, a one-man movement that made furniture from found objects such as car parts, kitchen pans and cast-iron railings. In 1987, he established the Dixon PID studio-workshop. By 1990 he was an internationally renowned design force and started designing lights. An excellent example of his individual approach is the pendant light on page 69, which is similar in form to a child's ball with a load of styrofoam cups stuck to it!

One charming thing about Tom is that he is far cleverer than he thinks he is, delivering originality again and again and again. Once the *enfant terrible* of the design scene, he is currently the design director of Habitat, that international, design-led, home-store emporium. His influence and canny vision will no doubt give the company a point of difference that will reposition it as a benchmark of affordable style and practicality. Innovation, daring, lateral thinking and charismatic style are all key ingredients of his work.

Left Star Lights, *designed in 1990 for Eurolounge. Here is a lightweight steel-rod construction covered with Japanese paper to enclose a conventional incandescent bulb.*

Right S Chair, *designed in 1991/92 for Cappellini. Tom Dixon does not limit his talents to lighting design. Here is one of his furniture pieces made from woven marsh reed over a mild-steel welded framework.*

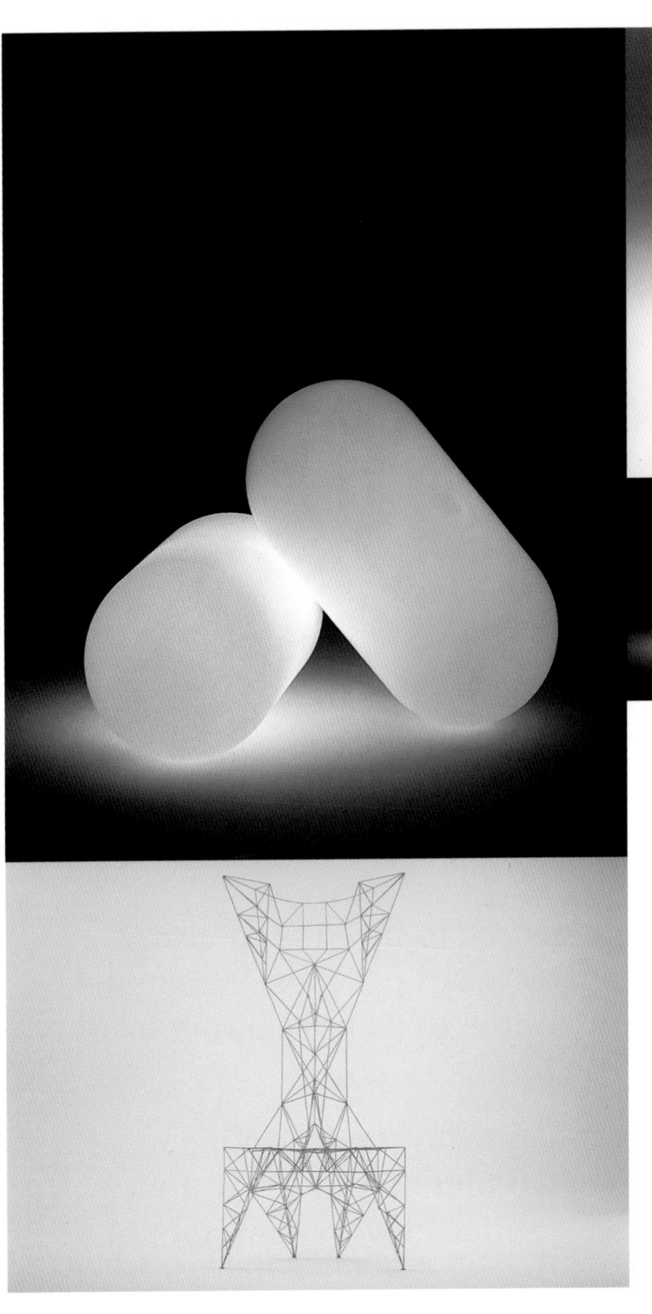

Above Bird, *designed in 1991 for Cappellini. Here is a padded chaise longue covered in wool to disguise its steel frame. It rocks a little when you first alight, but fortunately it quickly becomes reassuringly stable.*

Left Melon, *designed in 1997 for Eurolounge. Looking like giant pills, these luminous, phosphorescent, rotationally molded polyethylene capsules work well in multiples.*

Above Star *(see page 69).*

Left Pylon Chair, *designed in 1992 for Cappellini. Inspired by early "wire frame" computer drawings, this painted mild-steel rod construction chair is surprisingly comfortable.*

Right *More* Star Lights *(see opposite).*

Opposite Space Projector, *designed by the in-house team at Mathmos in 1999. The space projector base and body on this occasional light are made of ABS, while the focusing lens and slide carrier are made of nylon.*

fact file tips on buying lights

Planning is the key.

* Foresee any installation requirements. For example, some lighting requires transformers that need to be either hidden or contained, and many wall lights will need to have wiring fished into the wall and then covered over with sheetrock or plaster.
* Decide on whether you want the light source to be either a visible feature or camouflaged or hidden from view. Art galleries, for example, are mostly designed so that the light source does not detract from the artwork on the walls.
* Take into consideration whether you want, say, task lighting or decorative lighting; task lighting or ambient lighting; or any combination. Certain tasks require a standard level of illumination for safety reasons.

* Consider the reflective, directive and bouncing qualities of light. If you have mirrors behind your lighting, the strength of the bulb may not need to be as high. The color of the walls behind the lights will also affect the quality, since lighter colors will reflect and darker colors will absorb the light.
* Check the compatibility and availability of bulbs.
* Check the plug and outlet compatibility: different countries have different requirements.
* Plan to use dimmers whenever possible, since they add flexibility to your lighting plan. Dimmers for adjustable levels of ambient light in the bathroom are also useful. But make sure that the switches are placed outside the room to avoid fumbling for them in the dark.
* Dimmers reduce light but do not save much electricity, so use a lower wattage bulb.
* Install at least some of the light control systems by the main door so you don't have to stumble in or out of a dark room.
* Balance task lighting with ambient light to avoid shadows.
* Safety and stability are important. Freestanding uplighters can be hazardous when children and pets are around.

* Lights can also be fire hazards—items falling on or into bulbs can catch fire, causing an explosion.
* Take environmental concerns into account: save electricity by attaching timers or sensors to some key lights so that they turn off automatically when you are out of the home.
* The key differences between the various bulb options lie in their efficiency and the quality and quantity of light they emit. Here is a comparison of lamp efficiencies:

Description	*Average life*	*Temperature*	*Light output*	*Energy efficiency*
100W incandescent globe	1,000–2,000 hours	2700K	1,200 lumens	12 lumens/watt
500W quartz-halogen	3,000 hours	3000K	9,500 lumens	19 lumens/watt
65W fluorescent tube	9,000 hours	3000K	5,000 lumens	76 lumens/watt
18W compact fluorescent tube	10,000 hours	3000K	1,500 lumens	83 lumens/watt

The compact fluorescent tube is about seven times more energy efficient than the incandescent lamp, albeit at the expense of light quality.

suppliers & designers

The following stores stock a good range of contemporary lighting. If you are looking for a particular piece, it is worth telephoning the store first to make sure they still carry the line.

ARKITEKTURA SHOWROOMS
560 Ninth Street
San Francisco, CA 94103
Tel: (800) 400-4869
and
474 North Old Woodward Avenue
Birmingham, MI 48009
Tel: (800) 844-1126
Lighting and furniture by Artemide, Barovier & Toso, Driade, Flos, Fontana Arte, Ingo Maurer, Leucos, Luceplan and others

ATYS
306 South Main Street, 1D
Ann Arbor, MI 48104
Tel: (734) 996-2976
E-mail: mail@atys-inc
Web site: www.atys-inc.com
Lighting by Aero, Axis, Italiana Luce, Pallucco Italia, Philippe Starck, and others

CENTRO, INC.
4729 McPherson Avenue
St. Louis, MO 63108
Tel: (314) 454-0111
Products from Artemide, Fontana Arte, Foscarini, Italiana Luce, Leucos, Luceplan and Terzani

CHIMERA
319 A Street
Boston, MA 02210
Tel: (617) 542-3233
Showroom with products from Artemide, Babylon Design, Flos, Fontana Arte, Foscarini, Ingo Maurer, Italiana Luce, Leucos, Luceplan and others

CITY LIGHTS
1585 Folsom Street
San Francisco, CA 94103
Tel: (415) 863-2020
Lighting from Artemide, Flos and others; wide range of light bulbs

CONRAN SHOP
Michelin House
81 Fulham Road
London, N1 2TZ, U.K.
Tel: +44 207 589 7401
Products by Artemide, Fontana Arte, Foscarini, Luceplan, Paolo Rizzatto, Achille/Pier Giacomo Castiglioni, Flos, Philippe Starck, and others

CURRENT
1201 Western Avenue
Seattle, WA 98101
Tel: (206) 622-2433
Lighting from Aero and other contemporary designers

DDC-DOMUS DESIGN COLLECTION
181 Madison Avenue
New York, NY 10016
Tel: (212) 685-0800
E-mail: ddc@interport.net
Web site: www.ddc-newyork.com
Furniture and lighting by Vico Magistretti and other European contemporary designers

DESIGN CENTRO ITALIA
1290 Powell Street
Emeryville, CA 94608
Tel: (510) 420-0383
Web site: www.italydesign.com
Contemporary and modern lighting and furniture

DESIGN WITHIN REACH
Tel: (800) 944-2233 for catalog
Web site: www.dwr.com
Products by designers such as de Lucchi and Fassina, Meda and Rizzatto, Bellini, Breuer, Le Corbusier and Starck

Maglite flashlight

FORM AND FUNCTION
95 Vandam Street
New York, NY 10013
Tel: (212) 414-1800
Web site: www.info@
formandfunctiondesign.com
Vintage lighting and furniture from 1945 to 1975

FULL UPRIGHT POSITION
Tel: (800) 431-5134 for catalog
Web site: www.f-u-p.com
Modern lighting and furniture, including fixtures from Herman Miller and Louis Poulsen

ILLUMINATIONS
597 Cooledge Avenue
Atlanta, GA 30306
Tel: (404) 876-1064
Contemporary lighting from Artemide, Flos, Italiana Luce, Estaluz and others

ITALMODA
32968 Woodward Street
Royal Oak, MI 48073
Tel: (248) 549-1221
Lighting and furniture by European manufacturers and designers, including Foscarini

JULES SELTZER ASSOCIATES
8833 Beverly Boulevard
Los Angeles, CA 90048
Tel: (310) 274-7243
Web site: www.jules-seltzer.com
Lighting and furniture by Artemide, Flos, Vitra and others

LIMN CO.
290 Townsend Street
San Francisco, CA 94107
Tel: (415) 543-5466
and
501 Arden Way
Sacramento, CA 95815
Tel: (916) 564-2900
Web site: www.limn.com
Products by Artemide, ClassiCon, Driade, Fontana Arte, Pallucco Italia and others

LUMINAIRE
301 West Superior Street
Chicago, IL 60610
Tel: (800) 494-4358
and
7300 SW 45th Street
Miami, FL 33155
Tel: (305) 264-6308
Web site: www.luminaire.com
Lighting and furniture from ClassiCon, Flos, Fontana Arte, Ingo Maurer, Luceplan, Prandina and other manufacturers

THE MAGAZINE
1823 Eastshore Highway
Berkeley, CA 94710
Tel: (510) 549-2282
Web site: www.themagazine.org
Lighting and furniture by Aero, Arne Jacobsen, Le Corbusier, Herman Miller and others

MOBILI
2201 Wisconsin Avenue NW
Washington, DC 20007
Tel: (202) 337-2100
Lighting and furniture from Artemide, Carypen, Fontana Arte, Leucos, Ingo Maurer, Luceplan, Flos, Italiana Luce and others

MODERN LIVING
8775 Beverly Boulevard
Los Angeles, CA 90048
Tel: (310) 657-8775
E-mail:
modliv@modernprops.com
Web site:
www.modernliving.com
Lighting and furniture from designers such as Philippe Starck and Eileen Gray, and manufacturers including ClassiCon, Flos and Santa & Cole

MOMA DESIGN STORE
The Museum of Modern Art
44 West 53rd Street
New York, NY 10019
Tel: (800) 793-3167
Web site: www.momastore.org

THE MORSON COLLECTION
100 East Walton Street
Chicago, IL 60611
Tel: (800) 204-2514
and
31 St. James Avenue
Boston, MA 02116
Tel: (617) 482-2335
Contemporary lighting and furniture from Artemide, Leucos and Luceplan, and designers Eileen Gray and Antonio Citterio

MOSS
146 Greene Street
New York, NY 10012
Tel: (212) 226-2190
Lighting from Ingo Maurer and Flos

OK STORE
8303 West 3rd Street
Los Angeles, CA 90048
Tel: (323) 653-3501
E-mail: okstore@aol.com
Lighting from Aero and other manufacturers, lamps by Noguchi and Blake Simpson

O'VALÉ
2 Broad Street
Red Bank, NJ 07701
Tel: (732) 933-0437
Web site: www.redbank.com/
ovale/index.html
Products by Jacobsen, Starck, Henningsen, Foscarini, Louis Poulsen, Dansk Lighting, Italiana Luce and others

REPERTOIRE
114 Boylston Street
Boston, MA 02116
Tel: (617) 426-3865
E-mail: info@repertoire.com
Web site: www.repertoire.com
Lighting and furniture made by Cappellini, ClassiCon, Driade, Kartell, Luceplan and others

RETRO-MODERN
58 East 11th Street
New York, NY 10003
Tel: (212) 674-0530
Vintage Art Deco and Moderne light fixtures

ROOM
151 West 30th Street, Suite 705
New York, NY 10001
Tel: (800) 420-7666
or (212) 631-9900 for catalog
Web site: www.roomonline.com
Lighting from David Weeks, Julie Prisca, Casa Milano, DQ, Resolute and others

SEBASTIAN CONRAN ASSOCIATES
Design consultants
2 Munden Street
London W14 0RH U.K.
Tel: +44 207 602 0101

THE TERENCE CONRAN SHOP
344 East 59th Street
New York, NY 10022
Tel: (212) 755-9079
Lighting, furniture and accessories by the Conran design team and other designers

TOTEM
71 Franklin Street
New York, NY 10013
Tel: (888) 519-5587 for catalog
and West Coast location
E-mail: gail@totemdesign.com
Web site: www.totemdesign.com
Contemporary lighting from Aktiva, Axo, Box, Babylon Design, Colour Light Co., David Design, El Ultimo Grito, Eurolounge, Inflate, Jam and others

TRIOS HOME GALLERIE
1155 Canyon Boulevard
Boulder, CO 80302
Tel: (303) 442-8400
Web site: www.triosgallerie.com
Contemporary lighting and furniture from Italian manufacturers and designers

TROY
138 Greene Street
New York, NY 10012
Tel: (212) 941-4777
Lighting and furniture from Aero, Europa, Blake Simpson, Ou Baholyodhin, and others

VOLTAGE
2703 Observatory Avenue
Cincinnati, OH 45208
Tel: (513) 871-5483
Web site: www.voltageinc.com
Contemporary lighting from Artemide, Estaluz, Ingo Maurer, Italiana Luce, Mazzega, Resolute, Santa & Cole and Tobias Grau

Manufacturers

THE AMERICAN GLASS LIGHT COMPANY
979 Third Avenue (showroom)
New York, NY 10022
Tel: (212) 371-4800 for catalog
Light fixtures, including contemporary designs in the Art Deco and Moderne styles by Sandy Littman

ARTEMIDE
46 Greene Street
New York, NY 10013
Tel: (212) 925-1588
and
4200 Sepulveda Boulevard
Culver City, CA 90230
Tel: (310) 837-0179
and
208 W. Kinzie Street
Chicago, IL 60610
Tel: (312) 644-0510

BOND PROJECTS UK LTD
Prism Design Studio
38 Grosvenor Gardens
London SW1W OEB, U.K.
Tel: +44 207 730 3011

CAPPELLINI MODERN AGE
102 Wooster Street (showroom)
New York, NY 10012
Tel: (212) 966-0669

FLOS USA
200 McKay Road
Huntington Station, NY 11746
Tel: (800) 939-3567 (to be connected to nearest dealer)

FONTANA ARTE USA, INC.
8807 Beverly Boulevard
Los Angeles, CA 90048
Tel: (310) 247-9933
E-mail: fontanaarte@msn.com
Web site: www.fontanaarte.com

IL AMERICA INC.
109 Sanford Street
Handem, CT 06514
Tel: (203) 378-4000 for dealer locations
U.S. distributor for Italiana Luce and Foscarini

KARTELL
45 Greene Street
New York, NY 10012
Tel: (212) 966-6665
Web site: www.kartell.com
Furniture and lighting by Citterio, Magistretti, Starck and others

LAMPA
Tel: (516) 722-9450 for local dealers or catalog
Web site: www.lampa.com
Contemporary lighting and furniture

LEUCOS LIGHTING
P.O. Box 7829
Edison, NJ 08818
Tel: (800) 832-3360
Web site: www.leucos.com

LUCE INTERNATIONAL
300 Beale Street, Loft 415
San Francisco, CA 94105
Tel: (800) 591-3222
E-mail: luce@sirius.com
Manufacturers' agent for Flos, Leucos, Prisma, Resolute, Santa & Cole, and others, for contract design

LUCEPLAN USA INC.
315 Hudson Street
New York, NY 10013
Tel: (212) 989-6265 for local dealers
E-mail: luceplan@mail.idt.net
Web site: www.luceplan.com

MURANO DUE
3607 13th Avenue
Brooklyn, NY 11218
Tel: (718) 436-2002 for dealer locations or catalog
Lighting by designers Carlo Nason, Mauro Marzollo, Federico Codato and others

O LUCE
Web site: www.oluce.com

Auction houses

BUTTERFIELD & BUTTERFIELD
220 San Bruno Avenue
San Francisco, CA 94103
Tel: (415) 861-7500

CHRISTIE'S
502 Park Avenue
New York, NY 10022
Tel: (800) 247-4558

EBAY ONLINE AUCTIONS
Web site: www.ebay.com

SOTHEBY'S, INC.
1334 York Avenue
New York, NY 10021
Tel: (212) 606-7000

Further reading

Mel Byars: *50 Lights: Innovations in Design and Materials*
Pro Design Series, Whitney Library of Design, 1998.

Tom Dixon: *Architecture*
Design and Technology Press, 1990.

Phillippe Garner: *Sixties Design*
Benedikt Taschen Verlag GmbH, 1995.

Lynn Gordon: *ABC of Design*
Chronicle Books, 1996.

Jean Gorman: *Detailing Light*
Whitney Library of Design, 1995.

Thomas Hauffe: *Design*
Barron's Educational Series, 1996.

Wanda Jankowski:
Creative Lighting
Rizzoli Bookstore, 1997.

Peta Levi:
New British Design 1998
Mitchel Beazley, 1998.

Light Design
Stichting/Foundation, 1998.

Jeremy Myerson:
International Lighting Design
Books Nippan, 1996.

Pocket Design Directory
Janvier Publishing, 1998.

Starck
Benedikt Taschen Verlag GmbH, 2000.

Peter Tregenza and David Loe:
The Design of Lighting
E & FN Spon, 1998.

Janet Turner:
Designing with Light
Rotovision, 1999.

index

Figures in **bold** refer to photos/captions

First published 1999 by Conran Octopus Limited, a part of Octopus Publishing Group, London. North American edition published 2000 by Soma Books, by arrangement with Conran Octopus.

Soma Books is an imprint of Bay Books & Tapes,
555 De Haro St., No. 220, San Francisco, CA 94107.

For the Conran Octopus edition:
Commissioning Editor: Denny Hemming
Series Editor: Gillian Haslam
Project Editor: Emma Callery
Managing Editor: Kate Bell
Index: Emma Callery
Creative Director: Leslie Harrington
Art Editor: Lucy Gowans
Stylists: Emma Thomas and Sarah Hollywood
Production: Zoe Fawcett

For the Soma edition:
North American Editor: Karen O'Donnell Stein
Proofreader: Ken DellaPenta

Library of Congress Cataloging-in-Publication Data
Conran, Sebastian.
Soma basics—lighting / Sebastian Conran & Mark Bond ;
photography by Thomas Stewart.
p.cm.
Originally published: London : Conran Octopus, 1999.
Includes bibliographical references and index.
ISBN 1-57959-063-2 (pbk. : alk. paper)
1. Lighting. I. Bond, Mark. II. Title.
TH7703 .C634 2000
747'.92—dc21

99-054702

Printed in China
10 9 8 7 6 5 4 3 2 1

Distributed by Publishers Group West

Acknowledgments

Special thanks for Tim Gadd for research and for all his help behind the scenes. Thanks also to Peter Burian, Kathryn Mills and Alfred Munkenbeck.

The authors and publishers wish to thank the following for their considerable help and assistance: Josie Ballin at **Mathmos**; Gilla Bond at **Artemide**; Sharon Bowles at **Bowles and Linares**; Tamara Caspersz and James Mair at **Viaduct**; Clemente Cavigioli; Sophie Chandler; Norman Cull at **Number 16**; Fiona Dodd at **Same**; Joao Ferreira at **Proto UK**; Elena Graves at **Eurolounge**; Clementine Hope; Andrew Johnson and Gill Hicks at **Blueprint**; Frank Kelly at **MDS**; Joanne Leyland at **Purves & Purves**; Jeremy Lord; Fiona Mackenzie-Jenkin at **Inflate**; Michael Marriott; Alexis Nishihata at **Aero**; Dennis Ong at **Aktiva**; James Peto at **Design Museum**; Suzel Pitty at **Babylon**; Roy Sant; Vanessa Scheibner at **Tecta**; Jo Wolley at **Catalytico**.

With thanks for the following for the kind loan of transparencies: **Aero** (p.70 Paper Pendant, p.71 Anywhere Light), **Anglepoise** /www.anglepoise.co.uk (p.17), **Aram Design** (p.17 Gray), **Aktiva** (p.54 Filo, Swivel, p.71 Filo Pendant, p.78 Search Light SL-111, p.87 Icon "Holes" & Search Light), **Artemide** (p.17 Magistretti & Sapper, p.46 Iride & Latona, p.60 Shogun, p.79 Musa Sospensione, p.80 Mikado Track & Chronocolour Series, p.88 Acheo, p.93 Pantarei 300 Half Light), **Babylon** (p.75 UFL), **Box** (p.76 Waves), **Sophie Chandler** (p.57 Lightbowl 1, 2), **Sebastian Conran** (p.90 & 92), **David Design** (p.46 Lowlight and Highlight), **Tom Dixon** (p.118 portrait of Tom Dixon & Star Lights, p.119 Pylon Chair, Star Lights & Bird), **Flos** (p.17 Castiglioni, p.23 Fucsia, p.30 Wall Light, p.41, p.42 Taccia, p.46 Helice, p.55 Miss Sissi, p.71 Cina, p.72–73 Fucsia, p.78 Romeo Moon 52, p.83 Lastra 6, p.85 Ariette, p.86 Wall A Wall A & Train Train, p.87 Drop 2, p.110 Brera S, all of p.112–113, p.116 Miss Sissi, p.117 Lucifair, Light Lite & Ara), **Fontana Arte** (p.17 Chiesa, p.43 Elvis, p.50 "2198," p.55 "2198 TA," p.66 Ray, p.75 Alzaia), **Foscarini** (p.50 bottom left, p.88 Circus Grande), **Clementine Hope** (p.74 Light Light), **Inflate** (p.75 UFO), **Jeremy Lord** (p.86 Chromawall), **Pearson Lloyd** (p.55 Ilos), **Luceplan** (p.42 Costanza & Pod Lens, p.46 Lola floor & Lucilla, p.55 On Off, p.66 Fortebraccio, p.78 Titania, p.87 Lola Wall, p.88 Screen, p.93 Solar Bud, p.93 Pod Lens), **Mathmos** (p.60 Fibre Space, p.62 Lava Lamp, p.121 Space Projector), **Ingo Maurer** (p.14 Fly Candle Fly, p.75 Zuuk, p.80 Ya Ya Ho, p.89 Zero One, p.111 Oskar, all of p.114–115), **Jasper Morrison** (p.80 Glo-Ball Pendant) **Pallucco Italia** (p.17 Fortuny, p.47 Short Wave Long Wave), **Lucy Pope** (p.89 Rubber Light), **Proto UK** (p.70 Pendant Light), **Louis Poulsen** (p.17 Henningsen, p.29, p.67, p.81 PH Snowball & Satellite Pendant), **Purves & Purves** (p.117 Le Maire, Romeo Moon floor version), **Roy Sant** (p.71 Daisy), **Tecta** (p.17 Rietveld), **Nic Tompkin** (p.8 & 11).

Thanks for the following for the loan of accessories for photography: **Alma Leather** +44 207 375 0343, **Century Design** +44 207 487 5100, **Co-existence** +44 207 354 8817, **Grasslands & Savanah** +44 207 727 4727, **Mathmos** +44 207 549 2700, **Purves & Purves** +44 207 580 8223, **Robert Wyatt** +44 208 530 6891, **Same** +44 207 247 9992, **Sixty 6** +44 207 224 6066, **Shiu-Kay Kan** +44 207 434 4095, **Space** +44 207 229 6533, **Stepan Tertsakian** +44 207 236 8788, **Viaduct** +44 207 278 8456, **Walkahead** +44 207 275 8908.